T0224974

SpringerBriefs in Applied Sciences and Technology

SpringerBriefs present concise summaries of cutting-edge research and practical applications across a wide spectrum of fields. Featuring compact volumes of 50 to 125 pages, the series covers a range of content from professional to academic.

Typical publications can be:

- A timely report of state-of-the art methods
- An introduction to or a manual for the application of mathematical or computer techniques
- A bridge between new research results, as published in journal articles
- A snapshot of a hot or emerging topic
- An in-depth case study
- A presentation of core concepts that students must understand in order to make independent contributions

SpringerBriefs are characterized by fast, global electronic dissemination, standard publishing contracts, standardized manuscript preparation and formatting guidelines, and expedited production schedules.

On the one hand, **SpringerBriefs in Applied Sciences and Technology** are devoted to the publication of fundamentals and applications within the different classical engineering disciplines as well as in interdisciplinary fields that recently emerged between these areas. On the other hand, as the boundary separating fundamental research and applied technology is more and more dissolving, this series is particularly open to trans-disciplinary topics between fundamental science and engineering.

Indexed by EI-Compendex, SCOPUS and Springerlink.

Rabiu Muazu Musa · Anwar P. P. Abdul Majeed ·
Mohamad Razali Abdullah · Garry Kuan ·
Mohd Azraai Mohd Razman

Data Mining and Machine Learning in High-Performance Sport

Performance Analysis of On-field and Video Assistant Referees in European Soccer Leagues

 Springer

Rabiu Muazu Musa
Centre for Fundamental and Continuing
Education
Universiti Malaysia Terengganu
Kuala Nerus, Terengganu, Malaysia

Anwar P. P. Abdul Majeed
School of Robotics
XJTLU Entrepreneur College (Taicang)
Xi'an Jiaotong—Liverpool University
Suzhou, Jiangsu, China

Mohamad Razali Abdullah
East Coast Environmental Research
Institute
Universiti Sultan Zainal Abidin
Kuala Nerus, Terengganu, Malaysia

Garry Kuan
Exercise and Sports Science Programme
School of Health Sciences
Universiti Sains Malaysia
Kubang Kerian, Kelantan, Malaysia

Mohd Azraai Mohd Razman
Innovative Manufacturing
Mechatronics and Sports Laboratory
Faculty of Manufacturing and Mechatronic
Engineering Technology
Universiti Malaysia Pahang
Pekan, Pahang, Malaysia

ISSN 2191-530X   ISSN 2191-5318   (electronic)
SpringerBriefs in Applied Sciences and Technology
ISBN 978-981-19-7048-1   ISBN 978-981-19-7049-8   (eBook)
https://doi.org/10.1007/978-981-19-7049-8

This Springer imprint is published by the registered company Springer Nature Singapore Pte Ltd.
The registered company address is: 152 Beach Road, #21-01/04 Gateway East, Singapore 189721, Singapore

Acknowledgement

We would like to acknowledge Prof. Dr. Zahari Taha for the guidance as well as the valuable suggestions for making the realization of this book possible.

Rabiu Muazu Musa
Anwar P. P. Abdul Majeed
Mohamad Razali Abdullah
Garry Kuan
Mohd Azraai Mohd Razman

Contents

About the Authors

Dr. Rabiu Muazu Musa holds a Ph.D. degree in Sports Science from Universiti Sultan Zainal Abidin (UniSZA), Malaysia. He obtained his M.Sc. in Sports Science from UniSZA and his B.Sc. in Physical and Health Education at Bayero University Kano, Nigeria. His research activity focused on the development of multivariate and machine learning models for athletic performance. His research interests include sports performance analysis, health promotion, sport and exercise science, talent identification, test, and measurement as well as machine learning in sports. He obtained several research grants both as a principal and a co-investigator. He is an active reviewer to many reputable journals and currently serving as an academic editor for *PLOS ONE* as well as executive guest editor for the *Open Sports Science Journal* and *SN Applied Sciences*. He authored many publications in myriad journals, conference proceedings, books as well as chapters in books. He is currently a senior lecturer at the Centre for Fundamental and Continuing Education, Universiti Malaysia Terengganu.

Dr. Anwar P. P. Abdul Majeed graduated with a first-class honours B.Eng. in Mechanical Engineering from Universiti Teknologi MARA (UiTM), Malaysia. He obtained an M.Sc. in Nuclear Engineering from Imperial College London, UK. He then received his Ph.D. in Rehabilitation Robotics from the Universiti Malaysia Pahang (UMP). He is currently serving as an associate professor at the School of Robotics, XJTLU. Prior to joining XJTLU, he was a senior lecturer (an assistant professor) and was the head of Programme (Bachelor of Manufacturing Engineering Technology (Industrial Automation)) at the Faculty of Manufacturing and Mechatronics Engineering Technology, UMP. He is also currently serving as an adjunct lecturer at UCSI University, Malaysia. He is also a visiting research fellow at EUREKA Robotics Centre, Cardiff Metropolitan University, UK. He is a chartered engineer, registered with the Institution of Mechanical Engineers (IMechE), UK, a member of the Institution of Engineering and Technology (IET), UK, as well as a senior member of the Institute of Electrical and Electronics Engineers (IEEE). His research interest includes rehabilitation robotics, computational mechanics, applied mechanics, sports engineering, renewable and nuclear energy, sports performance analysis as well as machine learning. He has authored over 60 papers in different journals, conference proceedings as well as books. He serves as a reviewer in a number of prolific journals such as *IEEE Access, Frontiers in Bioengineering and Biotechnology, SN Applied Sciences, PeerJ Computer Science, Applied Computing and Informatics*, among others. He has also served as a guest editor for *SN Applied Sciences, MDPI, Frontiers*, as well as an editor for several Springer book series. He is currently serving as an academic editor for *PLOS ONE*, a review editor for *Frontiers in Robotics and AI* and a section editor for *Mekatronika* (UMP Press). He is also a member of the Young Scientists Network-Academy of Sciences Malaysia (YSN-ASM). With regard to learned/civil society activities, he is an active member of the IET Malaysia Local Network as well as acting as a liaison officer for the Imperial College Alumni Association Malaysia.

Assoc. Prof. Dr. Mohamad Razali Abdullah obtained his Bachelor of Physical Education in 1989 from Universiti Putra Malaysia (UPM). He obtained his M.Sc. in Sport and Exercise Science from the University of Wales Institute, Cardiff, in 1998, and in 2007, he received his Ph.D. in Sports Science from UPM. His research interests include motor control, sports biomechanics, motor performance, and machine learning in sports. He is currently an associate professor at East Coast Environmental Research Institute Universiti Sultan Zainal Abidin, Terengganu, Malaysia.

Assoc. Prof. Dr. Garry Kuan is an associate professor of the Exercise and Sports Science Programme, School of Health Sciences, Universiti Sains Malaysia, and a research fellow at the Brunel University, London, UK. He completed his Ph.D. in Sport Psychology at Victoria University, Melbourne, Australia. Presently, he is the secretary general of the Asian-South Pacific Association of Sport Psychology (ASPASP), the secretary general of the Malaysian Sport Psychology Association (MASPA), the executive board member of the Asian Council of Sports Science (ACESS), and the scientific committee of the World Exercise Medicine. He is the recipient of the 2021 International Society of Sport Psychology (ISSP) Developing Scholar Award, the 2021 Ten Outstanding Young Malaysian, the 2021 Asia Outstanding Academician of the year, the 2020 IFPEFSSA International Eminent Educator Award, A-CIPA Young Researcher Award at the 27th International Congress of Applied Psychology, International Scholars Award at the ICSEMIS pre-Olympic conference, and the Atsushi Fujita Research Scholarship at the 6th ASPASP conference. His research interest includes sport psychology, neuroscience, health promotion, test and measurements, and music. During his social time, he plays the first violin professionally and teaches communities to play various musical instruments.

Dr. Mohd Azraai Mohd Razman graduated his first degree from the University of Sheffield, UK, in Mechatronics Engineering. He then obtained his M.Eng. from Universiti Malaysia Pahang (UMP) in Mechatronics Engineering as well. He then completed his Ph.D. at UMP specifically in the application of machine learning in aquaculture. His research interest includes optimization techniques, control systems, signal processing, instrumentation in aquaculture, sports engineering as well as machine learning. He is currently serving as a guest editor for *SN Applied Sciences* in a number of topical collections. He has also edited a number of volume in Springer's *LNEE* and *AISC* series. He is currently serving as the editor-in-chief for *MEKA-TRONIKA: Journal of Mechatronics and Intelligent Manufacturing* under UMP Press.

Chapter 1
Current Trend of Analysis in High-Performance Sport and the Recent Updates in Data Mining and Machine Learning Application in Sports

1.1 An Overview of High-Performance Sport

The term "High Performance" can be referred to as a process that encapsulates the optimization of techniques and procedures to accomplish exceptional results. In the sporting context, sports teams and organizations prioritize high-performance culture [1]. A typical team or an organization of sport constitutes individuals with various backgrounds, abilities, and obligations. These unique sets of people require diligent leadership, thought, and action to thrive. Essentially, all the individuals involved need to work harmoniously to realize the goal of achieving excellence.

Because of the political and financial significance of elite athlete performance at both individual and team levels, many high-performing nations have included elite sports on their national policy agenda [2]. Many nations' policy agendas include an emphasis on elite sports policy, elite financing, and a deliberate approach to developing athletes. As a result of this, researchers are interested in better understanding elite sports systems, explaining variables that drive success and shaping policy.

The questions often asked are why some sports teams and nations excel while others fail in high-performance sports. The answer to this question has ignited debates which gave rise to the development of an emerging area of study for the past twenty years [2]. A few examples of such studies are reported by the preceding investigators [3–5]. However, it is important to note that several factors play a role in shaping performance at a high level, and such success in high-performance sports is determined by numerous indicators.

© The Author(s), under exclusive license to Springer Nature Singapore Pte Ltd. 2022
R. Muazu Musa et al., *Data Mining and Machine Learning in High-Performance Sport*,
SpringerBriefs in Applied Sciences and Technology,
https://doi.org/10.1007/978-981-19-7049-8_1

1.2   Responsibilities and Influence of Referees in Top-Flight Soccer Tournaments

The game of soccer also known as association football is considered an explosive team event in which players are required to execute intermittent and short duration of high-intensity exercises along with a concomitant recovery period of lower intensity [6–8]. It has been reported that at the top-flight soccer tournament, players cover distances exceeding 10 km per match which is approximately 600 m above 21 km per hour [9, 10]. As a result, professional soccer referees must be of a high level of fitness to follow the passage of the game throughout the players' rapid movements, changes of direction, and sprints. It is worth noting that referees must be in the correct area and at the exact moment during every action, and they must attentively observe players' behaviour, understand the rules, and make key judgments within seconds [11]. These demands and attributes make the work of referees highly complex and hitherto, saddled with the responsibility of keeping the game attractive via the implementation of valid and good judgement skills.

The referee is responsible for making split-second judgements, enforcing the game's laws on the participants, and presiding over the match from an unbiased point of view. However, his (or her) mistakes might have detrimental effects on clubs, fans, players, and teams from an economic, psychological, and social standpoint [12]. It is not uncommon to hear people insinuating after a soccer match when the game result typically did not favour them. Complaints such as "They were not better than us, the penalty awarded to them was a mistake, the referee was clearly biased as he ought to have awarded a penalty to us on a foul to Ronaldo. But today, they didn't even go to VAR when there were clear instances where the VAR should have been consulted" [13]. It is against this background that the introduction of the video assistant referee (VAR) was actualized to improve the decision-making of on-field referees. However, despite the integration of VAR into the current refereeing system, the performances of the referees are still questionable.

1.3   Recent Updates in Data Mining and Machine Learning Application in Sports

The employment of machine learning in sporting activities has gained significant attention in recent times. In fact, owing to its popularity, a number of review papers have been recently published with regard to this particular topic [14–20]. This section does not intend to compete with such literature, nonetheless, it aims to provide the readers with a flavour on the utilization of such technology in the sporting domain.

Tan et al. [21] investigated the efficacy of different machine learning models, i.e. k-nearest neighbours (kNN), support vector machine (SVM), artificial neural networks (ANN), Naive Bayes (NB), and random forest (RF) on the classification of different Speak Takraw kicks based on data captured via inertial measurement units

(IMU) placed on the shank of the participants. Different statistical features were extracted from the IMU data before it was fed into the aforesaid machine learning models. The 70:30 hold out ratio was utilized for training and testing, respectively. It was demonstrated from the study that based on the features extracted that the ANN model is capable of classifying well (no misclassification in the test dataset) the kicks investigated, viz. serve (a.k.a. "tekong"), feeder as well as spike.

Thabtah et al. [22] investigated the efficacy of machine learning and its associated features in predicting National Basketball Association (NBA) game outcome. NB, ANN, and decision tree (DT) were used in the study. The NBA Finals Team Stats Dataset with match outcomes from 1980 through 2017 was obtained from kaggle.com. Three different feature selection techniques were employed in the study, namely multiple regression, correlation feature set, and the RIPPER algorithm. It was indicated from the study that the defensive rebounds were found to be a significant feature in influencing the results of an NBA game.

The classification of gait and jump in modern equestrian was investigated by Echterhoff et al. [23]. A smartphone was attached to the horse saddle to acquire the accelerometry and gyroscopic data. A total of 268 different features were extracted from both the time and frequency domain, and a total of 22 variation of machine learning models were investigated by considering the ninefold cross-validation technique. It was shown from the study that the cubic-SVM could achieve a classification accuracy of 95.4% in distinguishing the four classes evaluated, namely walk, trot, canter, and jump.

McGrath et al. [24] investigated the detection of cricket fast bowling by means of machine learning from IMU acquired data. Different machine learning models were evaluated, namely linear SVM (LSVM), polynomial SVM, ANN, RF, and gradient boosting (XGB) from the reduced feature set of 223. Different sampling frequencies were also evaluated to investigate the efficacy of the developed models in classifying bowling and non-bowling events. It was illustrated from the study that the SVM models demonstrated a slightly higher classification accuracy against the other models evaluated in general although it should be noted that all models achieved an accuracy of more than 95%, suggesting that the features extracted were indeed significant.

The classification of boxing punches by means of machine learning via IMU extracted data was investigated by Worsey et al. [25]. The participant was instructed to perform jab, cross, hook, and uppercut using both the right and left hand. The principal component analysis (PCA) was used to reduce the dimensionality of the extracted features. A total of six machine learning models were evaluated, namely the logistic regression (LR), LSVM, Gaussian SVM (GSVM), ANN, RF, and XGB. It is worth noting that the hyperparameters of the models were evaluated by means of the exhaustive grid-search technique via fivefold cross-validation. It was demonstrated from the study that the untuned GSVM and the ANN model could classify the punches well.

A feature selection investigation was carried out by Duki et al. [26] in the classification of Taekwondo kicks. Nine statistical features were extracted from the acceleration data of an IMU, viz. minimum, maximum, mean, median, standard deviation, variance, skewness, kurtosis, and standard mean error. The significance of the features was investigated by means of ANOVA as well as chi-squares (χ^2) test, before it was fed into different machine learning models, namely SVM, RF, kNN, and NB, respectively. The 60:40 hold out ratio was used in the study. It was shown from the study that the features identified via the ANOVA method could yield a better classification accuracy up to 86.7% via the RF model in comparison with utilizing all features, or the features identified by χ^2.

Abdullah et al. [27] employed the use of six transfer learning models with optimized SVM classifier to classify skateboarding tricks. Two input image types that were extracted from an IMU that is fixed on the skateboard and transformed to stacked raw signals (RAW), and continuous wavelet transform (CWT) was investigated. It is worth noting that the hyperparameters of the SVM model were optimized via the grid-search technique, and a training, testing, and validation ration of 60:20:20 was used in the study. It was established from the study that the CWT-MobileNet-optimized SVM pipeline was determined to be the best among all the permutation considered by considering its computational time as well as its classification accuracy.

The identification of high-performance volleyball players (HVP) and low-performance volleyball players (LVP) based on anthropometric variables and psychological readiness indicators by means of machine learning was investigated by Musa et al. [28]. The Louvain clustering algorithm was used to distinguish the performance of the players, while the logistic regression model was utilized to classify the classes. Owing to the skewed nature of the data, the authors employed the Synthetic Minority Oversampling TEchnique (SMOTE) to artificially increase the minority class dataset to avoid the overfitting notion upon classification. It was shown from the study that an excellent identification of the class of the player could be attained based on the pipeline developed. It is apparent from the brief survey presented that machine learning has gained significant traction in the sporting domain and has demonstrated to be able to yield reasonably good prediction in a myriad of cases, suggesting its invaluable contribution towards sports in general.

1.4 Mann–Whitney U-Test and Kruskal–Wallis Analysis

The Mann–Whitney U-test is a univariate mathematical analysis that is mostly used to compare means. The U-test is a type of dependency test analysis that assumes the variables in the analysis may be classified as independent or dependent [29]. The test's logic is based on the assumption that fluctuations in the average scores of the dependent variable(s) are mostly attributable to the independent variable's direct influence(s). It is worth mentioning that the independent variable(s) is also known as a factor since it splits the observed samples into two or more clusters.

The Kruskal–Wallis test, developed by Kruskal and Wallis in 1952, is a nonparametric method for determining whether samples are drawn from the same distribution. It broadens the Mann–Whitney U-test to include more than two groups [30, 31]. The Kruskal–Wallis test's null hypothesis is that the mean rankings of the groups are the same. Kruskal–Wallis's test is known as one-way ANOVA on ranks because it is the nonparametric counterpart of one-way ANOVA [32].

As opposed to the *t*-test and *F*-test, the Mann–Whitney *U*-test and Kruskal–Wallis's test are considered a nonparametric analysis which signifies that no prior assumption is established on the means of the distribution of the sample within the variables of interests in the population samples. Unlike the equivalent one-way ANOVA, the nonparametric Mann–Whitney *U*-test and Kruskal–Wallis's test do not assume that the underlying data is normally distributed [33].

These tests have been successfully applied in different sports and have been shown to be effective in projecting differences between two or more levels of performance classes. For instance, in a recent study, Mann–Whitney *U*-test has been applied to identify the technical as well as tactical performance indicators that could differentiate between the successful and unsuccessful teams in elite beach soccer competitions [34]. In an earlier study, the Mann–Whitney *U*-test was also successfully employed to study the probability of sustaining sports injuries among British athletes partaking in wheelchair racing [35]. On the other hand, Kruskal–Wallis's test was reported to be effective in comparing and separating the use of substances in ballet, dance sport, and synchronized swimming [36]. Similarly, the test has proved useful in ascertaining the effect of open and closed-skill sports on the cognitive functions of amateur table tennis athletes [37].

1.5 Features Extraction Analysis via Information Gain

In machine learning, information gain (IG) is a typical entropy-based function assessment approach. The IG approach is often used to extract information from one or more features concerning a particular categorical-dependent variable [38]. It should be noted that in the current study, IG is used to assess the functionality that may be employed in delivering information in order to estimate the significance of a certain variable for classification or discrimination tasks. The IG is used in the current investigation to extract information that demonstrates the importance of the functions, i.e. the performance parameters, in explaining the underlying associations with the performance of referees in this sport.

1.6 Cluster Analysis

1.6.1 Hierarchical Agglomerative Cluster Analysis (HACA)

Hierarchical agglomerative cluster analysis is commonly used as a tool of exploration as well as a non-exploratory method by which a cluster hierarchy for a single observation is established and a set of related observations form a distinct observation [39].

It is important to note that in this algorithm, the learning process is decided by the merges as well as the splits of the dataset, which are also implemented to isolate and illustrate identical findings in a dendrogram [7, 40, 41]. It should be remembered that HACA shows the number of clusters dependent on the vicinity of a given or predetermined cluster in the dendrogram. Distance of cosine was used in this analysis, and the clustering validation technique was conducted by class centroids [42].

1.6.2 Louvain Clustering

The Louvain clustering method is also regarded as the most recent clustering algorithm capable of classifying a given collection of data or observations. The method is designed to complete the work in two discrete parts; in the first, it seeks for a "thin" group by maximizing modularity in a classical approach. In the second stage, the algorithm connects nodes from related communities to form a distinct community, resulting in the formation of a new network of community nodes [43]. These steps can be repeated repeatedly until a modularity condition is met. This step also adds to the system's hierarchical fragmentation and the production of many divisions [44]. The divisions are often based on the density of the communities' borders, rather than the intercommunity margins.

1.6.3 K-means Cluster Analysis

A k-means clustering method is a form of cluster analysis approach that divides a set of data into k-predefined and non-overlapping subgroups called clusters, with only one group given to each data point [40, 45, 46]. The method seeks to make the inter-cluster data points as connected as feasible while maintaining the intra-cluster data points as different as possible. The cluster analysis was used to assign into groups based on performance indicators evaluated in this study. It is worth noting that the Euclidean distance was used as a distance metric to assign the formation of all clusters established in the study.

1.7   Principal Component Analysis as Data Mining Technique

PCA is a mathematical method used primarily to identify the structure of a dataset from a group of observed variables [40, 45]. PCA provides information about key variables that might reflect a particular dataset by observing the spatial and temporal heterogeneity of the entire dataset. The process of extracting information from the

PCA is performed by removing the data that is made up of the least important component and subsequently retaining the most useful information in the data [41, 47]. The employment of PCA is non-trivial in removing the most important information from a large dataset, which is crucial, as the analysis may serve to avoid wasting effort, cost, and time since the original data is often retained.

1.8 Application of Machine Learning Models in the Study

In this brief, different supervised machine learning models, namely *k*-nearest neighbours (*k*NN), support vector machine (SVM), logistic regression (LR), and artificial neural networks (ANN), were utilized towards the classifying different classes investigated. It is worth noting that the hyperparameters of the models evaluated were at times optimized, and such instances are explicitly mentioned in the ensuing chapters that involves the employment of machine learning models. The readers are encouraged to refer to our previous works [48–51], on the definition of the aforesaid machine learning models as well as performance indices that are often used to evaluate the efficacy of developed machine learning pipeline.

1.9 Datasets for the Study

A total of 6232 matches from five consecutive seasons (2017 through 2022) officiated across the English Premier League, Spanish Laliga, Italian Serie A, French Ligue1 as well German Bundesliga were retrieved from InStat Scout. The InStat Scout is one of the leading sports performance analysis companies founded in Moscow in 2007 with currently over 900 offices globally. The data is made available for the users upon subscription. InStat reported that the company examines referee performance in greater depth than any other statistics firm. InStat Scout includes a profile for each referee and generates a unique report after each match.

Due to the relative importance of referees' statistics for both the national federation and referees themselves, InStat developed and covered over 63 indicators for each referee, and to ensure the reliability and validity of the indicators, each indicator is linked to videos that provide the activity profile of the referee as well as comprehensive reports on the referee's overall activity in the match [52]. The indicators were developed after a thorough consultation with top referees from numerous countries. These indicators are from a complete profile of the referee's actions during a match. They covered aspects that constitute details statistics, decisions making, home and away teams' performance aggregates, injury time, distance metrics as well as reasons for awarding cards to players. These statistics are non-trivial as they portray the extent to which referees influence a match and how well they can officiate and keep up with the game. Table 1.1 depicts a detailed description of the datasets utilized in the study.

Table 1.1 Full datasets description in the study

Main statistics	Disputable decisions	Home-away teams aggregate
Matches refereed	Decisions which led to disputes	Home team goals
Fouls	Foul not detected	Away team goals
Challenges	Offside not detected	Fouls-home team
Yellow cards shown	Yellow cards (not shown)	Fouls-away team
Red cards shown	Red card (not shown)	Yellow cards (home team)
Direct red cards	VAR decision	Yellow cards (away team)
Red cards for two yellow cards	Goal disallowed	Red cards (home team)
Fouls per card		Red cards (away team)
Booking fouls (yellow card)		Booking fouls (home team)
		Booking fouls (away team)
Injury time	Distance metrics	Cards reasons
Injury time	Average distance to event	Air challenges
First half injury time	Distance to fouls	Ground challenges
Second half injury time	Distance to fouls leading to cards	Handball
Ball in play	Distance to goals	Challenges off the ball
% of actual ball in play time	Avg. referee distance to penalty box + 2 m around fouls	Dangerous play
Remaining time	Decisions in key zones, distance to 8 m	Misconduct
% of actual non-played time	Decisions in key zones, distance to 8–20 m	Simulation/diving
	Decisions in key zones, distance more than 2 m	Attack wrecking
	Average referee distance in other zones	Professional foul
	Decisions in other zones from close range (less than 25 m)	
	Decisions in other zones from a distance (more than 25 m)	

References

1. B. Houlihan, Commercial, political, social and cultural factors impacting on the management of high performance sport, in *Managing High Performance Sport* (Routledge, 2013), pp. 49–61
2. P. Sotiriadou, V. De Bosscher, Managing high-performance sport: introduction to past, present and future considerations (2018)
3. S.S. Andersen, L.T. Ronglan, *Nordic Elite Sport: Same Ambitions, Different Tracks* (Copenhagen Business School Press DK, 2012)
4. N.A. Bergsgard, B. Houlihan, P. Mangset, S.I. Nødland, H. Rommetvedt, *Sport Policy* (Routledge, 2009)

5. V. De Bosscher, P. De Knop, M. Van Bottenburg, S. Shibli, A conceptual framework for analysing sports policy factors leading to international sporting success. Eur. Sport Manag. Q. **6**, 185–215 (2006)
6. M.R. Abdullah, R.M. Musa, A.B.H.M. Maliki, N.A. Kosni, P.K. Suppiah, Development of tablet application based notational analysis system and the establishment of its reliability in soccer. J. Phys. Educ. Sport **16**, 951–956 (2016). https://doi.org/10.7752/jpes.2016.03150
7. A.B.H.M. Maliki, M.R. Abdullah, H. Juahir, F. Abdullah, N.A.S. Abdullah, R.M. Musa, S.M. Mat-Rasid, A. Adnan, N.A. Kosni, W.S.A.W. Muhamad, N.A.M. Nasir, A multilateral modelling of youth soccer performance index (YSPI). IOP Conf. Ser. Mater. Sci. Eng. **342**, 012057 (2018). https://doi.org/10.1088/1757-899X/342/1/012057
8. B. Drust, T. Reilly, N.T. Cable, Physiological responses to laboratory-based soccer-specific intermittent and continuous exercise. J. Sports Sci. **18**, 885–892 (2000)
9. M. Weston, C. Castagna, W. Helsen, F. Impellizzeri, Relationships among field-test measures and physical match performance in elite-standard soccer referees. J. Sports Sci. **27**, 1177–1184 (2009)
10. P. Krustrup, W. Helsen, M.B. Randers, J.F. Christensen, C. Macdonald, A.N. Rebelo, J. Bangsbo, Activity profile and physical demands of football referees and assistant referees in international games. J. Sports Sci. **27**, 1167–1176 (2009)
11. V. Fernández-Ruiz, Á. López-Samanes, J. Del Coso, J. Pino-Ortega, J. Sánchez-Sánchez, P. Terrón-Manrique, M. Beato, V. Moreno-Pérez, Influence of football match-play on isometric knee flexion strength and passive hip flexion range of motion in football referees and assistant referees. Int. J. Environ. Res. Public Health **18**, 11941 (2021)
12. A. Guillén, Exploración de indicadores para la medición operativa del concepto del Buen Vivir (2016)
13. R.D. Samuel, G. Tenenbaum, Y. Galily, An integrated conceptual framework of decision-making in soccer refereeing. Int. J. Sport Exerc. Psychol. **19**, 738–760 (2021)
14. F. Hammes, A. Hagg, A. Asteroth, D. Link, Artificial intelligence in elite sports—a narrative review of success stories and challenges. Front. Sports Act. Living **4** (2022)
15. A. Rossi, L. Pappalardo, P. Cintia, A narrative review for a machine learning application in sports: an example based on injury forecasting in soccer. Sports **10**, 5 (2021)
16. H. Van Eetvelde, L.D. Mendonça, C. Ley, R. Seil, T. Tischer, Machine learning methods in sport injury prediction and prevention: a systematic review. J. Exp. Orthop. **8**, 27 (2021). https://doi.org/10.1186/s40634-021-00346-x
17. R. Muazu Musa, A.P.P. Abdul Majeed, M.Z. Suhaimi, M.A. Mohd Razman, M.R. Abdullah, N.A. Abu Osman, Nature of volleyball sport, performance analysis in volleyball, and the recent advances of machine learning application in sports, in *Machine Learning in Elite Volleyball* (2021), pp. 1–11
18. M. Herold, F. Goes, S. Nopp, P. Bauer, C. Thompson, T. Meyer, Machine learning in men's professional football: current applications and future directions for improving attacking play. Int. J. Sports Sci. Coach. **14**, 798–817 (2019)
19. J.G. Claudino, D. de Oliveira Capanema, T.V. de Souza, J.C. Serrão, A.C. Machado Pereira, G.P. Nassis, Current approaches to the use of artificial intelligence for injury risk assessment and performance prediction in team sports: a systematic review. Sports Med. **5**, 1–12 (2019)
20. C. Richter, M. O'Reilly, E. Delahunt, Machine learning in sports science: challenges and opportunities. Sports Biomech. 1–7 (2021)
21. F.Y. Tan, M.H.A. Hassan, A.P.P. Abdul Majeed, M.A. Mohd Razman, M.A. Abdullah, Classification of Sepak Takraw kicks using machine learning, in *Human-Centered Technology for a Better Tomorrow* (Springer, 2022), pp. 321–331
22. F. Thabtah, L. Zhang, N. Abdelhamid, NBA game result prediction using feature analysis and machine learning. Ann. Data Sci. **6**, 103–116 (2019)
23. J.M. Echterhoff, J. Haladjian, B. Brügge, Gait and jump classification in modern equestrian sports, in *Proceedings of the 2018 ACM International Symposium on Wearable Computers* (2018), pp. 88–91

24. J.W. McGrath, J. Neville, T. Stewart, J. Cronin, Cricket fast bowling detection in a training setting using an inertial measurement unit and machine learning. J. Sports Sci. **37**, 1220–1226 (2019)

25. M.T.O. Worsey, H.G. Espinosa, J.B. Shepherd, D.V. Thiel, An evaluation of wearable inertial sensor configuration and supervised machine learning models for automatic punch classification in boxing. IoT **1**, 360–381 (2020)

26. M.S.M. Duki, M.N.A. Shapiee, M.A. Abdullah, I.M. Khairuddin, M.A.M. Razman, A.P.P.A. Majeed, The classification of Taekwondo kicks via machine learning: a feature selection investigation. MEKATRONIKA **3**, 61–67 (2021)

27. M.A. Abdullah, M.A.R. Ibrahim, M.N.A. Shapiee, M.A. Zakaria, M.A.M. Razman, R.M. Musa, N.A.A. Osman, A.P.P.A. Majeed, The classification of skateboarding tricks via transfer learning pipelines. PeerJ Comput. Sci. **7**, e680 (2021)

28. R.M. Musa, A.P.P. Abdul Majeed, M.Z. Suhaimi, M.R. Abdullah, M.A. Mohd Razman, D. Abdelhakim, N.A. Abu Osman, Identification of high-performance volleyball players from anthropometric variables and psychological readiness: a machine-learning approach. Proc. Inst. Mech. Eng. Part P J. Sports Eng. Technol. 17543371211045452 (2021)

29. R. Muazu Musa, A.P.P. Abdul Majeed, M.R. Abdullah, A.F. Ab. Nasir, M.H. Arif Hassan, M.A. Mohd Razman, Technical and tactical performance indicators discriminating winning and losing team in elite Asian beach soccer tournament. PLoS ONE **14**, e0219138 (2019)

30. Y. Xia, Correlation and association analyses in microbiome study integrating multiomics in health and disease. Prog. Mol. Biol. Transl. Sci. **171**, 309–491 (2020)

31. P.E. McKight, J. Najab, Kruskal–Wallis test, in *The Corsini Encyclopedia of Psychology 1* (2010)

32. M.A. Gipit, M.R.A. Charles, R.M. Musa, N.A. Kosni, A.B.H.M. Maliki, The effectiveness of traditional games intervention programme in the improvement of form one school-age children's motor skills related performance components (2017)

33. T.W. MacFarland, J.M. Yates, Mann–Whitney U test, in *Introduction to Nonparametric Statistics for the Biological Sciences Using R* (Springer, 2016), pp. 103–132

34. R.M. Musa, A.P.P.A. Majeed, N.A. Kosni, M.R. Abdullah, Technical and tactical performance indicators determining successful and unsuccessful team in elite beach soccer, in *Machine Learning in Team Sports* (Springer, 2020), pp. 21–28

35. D. Taylor, T. Williams, Sports injuries in athletes with disabilities: wheelchair racing. Spinal Cord **33**, 296–299 (1995)

36. B. Novosel, D. Sekulic, M. Peric, M. Kondric, P. Zaletel, Injury occurrence and return to dance in professional ballet: prospective analysis of specific correlates. Int. J. Environ. Res. Public Health **16**, 765 (2019)

37. S. Pancar, The effect of open and closed-skill sports on cognitive functions. Sport. Bakış Spor Eğitim Bilim. Derg. **7**, 159–166 (2020)

38. R.M. Musa, A.P.P. Abdul Majeed, A. Musa, M.R. Abdullah, N.A. Kosni, M.A.M. Razman, An information gain and hierarchical agglomerative clustering analysis in identifying key performance parameters in elite beach soccer (2021). https://doi.org/10.1007/978-981-15-7309-5_26

39. O. Maimon, L. Rokach, *Data Mining and Knowledge Discovery Handbook* (2005). https://doi.org/10.1007/b107408

40. M.R. Razali, N. Alias, A. Maliki, R.M. Musa, L.A. Kosni, H. Juahir, Unsupervised pattern recognition of physical fitness related performance parameters among Terengganu youth female field hockey players. Int. J. Adv. Sci. Eng. Inf. Technol. **7**, 100–105 (2017)

41. R.M. Musa, M.R. Abdullah, A.B.H.M. Maliki, N.A. Kosni, S.M. Mat-Rasid, A. Adnan, H. Juahir, Supervised pattern recognition of archers' relative psychological coping skills as a component for a better archery performance. J. Fundam. Appl. Sci. **10**, 467–484 (2018)

42. R. Muazu Musa, A.P.P. Abdul Majeed, Z. Taha, M.R. Abdullah, A.B. Husin Musawi Maliki, N. Azura Kosni, The application of artificial neural network and k-nearest neighbour classification models in the scouting of high-performance archers from a selected fitness and motor skill performance parameters. Sci. Sports (2019). https://doi.org/10.1016/j.scispo.2019.02.006

43. C. Wu, R.C. Gudivada, B.J. Aronow, A.G. Jegga, Computational drug repositioning through heterogeneous network clustering. BMC Syst. Biol. **7**, S6 (2013). https://doi.org/10.1186/1752-0509-7-S5-S6

44. V.D. Blondel, J. Guillaume, R. Lambiotte, E. Lefebvre, Fast unfolding of community hierarchies in large networks. J. Stat. Mech. Theory Exp. P10008 (2008). https://doi.org/10.1088/1742-5468/2008/10/P10008

45. H. Azahari, H. Juahir, M.R. Abdullah, R.M. Musa, V. Eswaramoorthi, N. Alias, S.M. Mat-Rashid, N.A. Kosni, A.B.H.M. Maliki, N.B. Raj, A multivariate analysis of cardiopulmonary parameters in archery performance. Hum. Mov. **19**, 35–41 (2019). https://doi.org/10.5114/hm.2018.77322

46. Z. Taha, M. Haque, R.M. Musa, M.R. Abdullah, A. Maliki, N. Alias, N.A. Kosni, Intelligent prediction of suitable physical characteristics toward archery performance using multivariate techniques. J. Glob. Pharma Technol. **9**, 44–52 (2009)

47. R.M. Musa, M.R. Abdullah, A.B.H.M. Maliki, N.A. Kosni, M. Haque, The application of principal components analysis to recognize essential physical fitness components among youth development archers of Terengganu, Malaysia. Indian J. Sci. Technol. **9** (2016)

48. R.M. Musa, A.P.P.A. Majeed, N.A. Kosni, M.R. Abdullah, *Machine Learning in Team Sports: Performance Analysis and Talent Identification in Beach Soccer & Sepak-Takraw* (Springer Nature, 2020)

49. R.M. Musa, A.P.P.A. Majeed, M.Z. Suhaimi, M.A.M. Razman, M.R. Abdullah, N.A.A. Osman, *Machine Learning in Elite Volleyball: Integrating Performance Analysis, Competition and Training Strategies* (Springer, 2021)

50. R. Muazu Musa, Z. Taha, A.P.P. Abdul Majeed, M.R. Abdullah, *Machine Learning in Sports* (Springer Singapore, Singapore, 2019). https://doi.org/10.1007/978-981-13-2592-2

51. M.A.M. Razman, A.P.A. Majeed, R.M. Musa, Z. Taha, G.-A. Susto, Y. Mukai, *Machine Learning in Aquaculture Hunger Classification* (Springer, 2020)

52. INSTAT FOR REFEREES—InStat

Chapter 2
Pattern Recognition of Misconducts Offences and Bookings of Top European Soccer Leagues Referees

2.1 Overview

Fouls and misconduct are acts made by players that are judged to be unfair by the referee resulting in punishment by issuing cards or verbal warning during a soccer match. Depending on the nature of the offence and the circumstances surrounding it, an offence may be classified as a foul, misconduct, or both. Law 12 in the Laws of the Game has addressed the issues of misconducts and fouls of the game [1]. A clear distinction has been drawn between actions related to fouls and actions that are deemed to be misconduct. A foul is an unfair action committed by a player that is regarded by the referee to violate the game's regulations and interfere with the active play of the game. Fouls are penalized by awarding the other side a free kick or a penalty kick depending upon where the offence is committed on the pitch. Misconduct, on the other hand, is defined as any behaviour by a player that the referee believes warrants a disciplinary consequence (warning or dismissal). Misconduct may involve behaviours that are also illegal. Misconduct, unlike fouls, can occur at any moment, even when the ball is out of play, at half-time, and before and after the game, and both players and substitutes can be penalized for misconducts actions. Misconduct may result in the player getting a warning by indicating a yellow card or being expelled from the pitch by showing a red card [2].

Referees are saddled with the responsibilities of officiating the match truthfully without taking any side. In a game of professional soccer, referees are assigned to regulate matches under the Laws of Association Football, which are defined by the governing body of world soccer, FIFA. In the European championship, member organizations and clubs, coupled with their players, officials, and members, as well as any person delegated by UEFA to carry out a role, must adhere to the Laws of the Game as well as UEFA's Statutes, regulations, directives, and decisions, and must conform with the UEFA Code of Ethics with ethical ideals, loyalty, integrity, and sportsmanship [1, 3].

© The Author(s), under exclusive license to Springer Nature Singapore Pte Ltd. 2022
R. Muazu Musa et al., *Data Mining and Machine Learning in High-Performance Sport*,
SpringerBriefs in Applied Sciences and Technology,
https://doi.org/10.1007/978-981-19-7049-8_2

The referee has extensive authority in enforcing the laws, including evaluating whether conduct is cautionable offence within its broad categories. As a result, referees' decisions can be contentious. Some laws may stipulate the conditions in which a warning should or must be issued, and several guidelines to referees provide additional advice. Nonetheless, the referees are encouraged to utilize their common sense in the provision of judgement [2]. For this and many other reasons, soccer referees are often accused of partial decision-making, especially in the aspect of issuing warnings or dismissals of players. It is essential, therefore, to investigate the activity of the top European referees in dealing with and issuing cautions to players based on different types of offences committed by the players. Thus, the present investigation endeavour is to identify the most prevalent actions of the players that frequently warrant the issuance of cautions or dismissals.

2.2 Data Treatment

The dataset containing 6232 matches from five consecutive seasons across the five top European leagues, i.e. the English Premier League, Spanish Laliga, Italian Serie A, French Ligue1 as well as German Bundesliga, was utilized for this study. These datasets included activities of the referees concerning the issuance of cautions in form of yellow cards as well as dismissals of the players through the issuing of red cards based on various actions related to misconducts and fouls. It is worth noting that before the commencement of the full analysis in this study, the data was preprocessed and checked for any missing information; rows that contained missing information are removed from the dataset [4, 5].

2.3 Principal Component Analysis for Future Selection

In this study, a principal component analysis (PCA) was applied to recognize the patterns of disciplinary actions undertaken by the referees via the extraction of the most relevant actions [6–8]. It is worth stating that the process of extracting information from the PCA is performed by removing the data that is made up of the least important component and subsequently retaining the most useful information in the data [9, 10]. The statistical analysis was executed via XLSTAT2014 add-in software for Windows.

2.4 Results and Discussion

Figure 2.1 shows the scree plot of the Eigenvalues for the PCA analysis. It could be observed from the figure that the PCA demonstrated a total of two components that

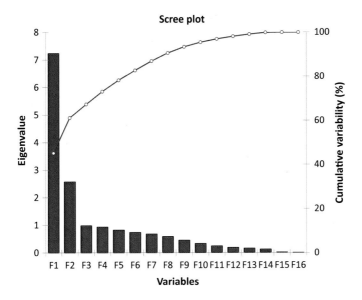

Fig. 2.1 Eigenvalue of the preliminary PCA analysis for components extractions

are highly attributed to the activities of the referees. In each of these two components, some specific actions and misconducts are identified and considered as most prevalent due to their relatively higher Eigenvalues (greater than 1). These identified components were retained and subsequently used as inputs parameters for further analysis, i.e. varimax rotation.

Table 2.1 tabulates the PCA analysis after varimax rotation. It could be seen from the table that in each of the identified two components, some relevant related fouls and misconducts actions are apparent. These indicators are identified due to satisfying the preset factor loading threshold, i.e. greater, or equal to 0.70. Likewise, it could be observed that a total number of 10 foul-related actions and misconducts, out of the 16 initially examined, were identified within the two components as most pronounced in the top European leagues' tournaments.

The findings from the present investigation revealed that there are high occurrences of fouls and misconducts-related actions in the top European soccer leagues. In PC1, the activities of the referees are heightened by fouls, challenges, challenges per foul, yellow cards, and fouls per card (most fouls committed attract cards). In principal component 2, air challenges, ground challenges, dangerous play, misconduct, as well as attack wrecking are found to be highly committed (Table 2.1). These variables projected disciplinary-related issues, aggressive plays, and the team's tactical errors which led to the issuance of warnings or cards. Data from a previous study has shown that aggressive plays have substantially increased in the Premier League alone with referees issuing a total of 300 red cards which is 60 per year and translates to approximately one card per every six games [11]. The data has further indicated that the most successful teams in the league are more likely to adopt aggressive plays. It

Table 2.1 Principal component analysis results after rotation

Indicators	PC1	PC2
Fouls	0.922	
Challenges	0.975	
Challenges per foul	0.944	
Yellow cards	0.822	
Direct red cards		
RC for two YC		
Fouls per card	0.929	
Air challenges		0.742
Ground challenges		0.826
Handball		
Challenge off the ball		
Dangerous play		0.802
Misconduct		0.828
Simulation/diving		
Attack wrecking		0.826
Professional foul		
Eigenvalue	**7.234**	**2.569**
Variability (%)	**45.212**	**16.055**
Cumulative (%)	**45.212**	**61.267**

has also been reported that a less powerful team is more likely to engage in damaging behaviours towards an opponent. On the other hand, disagreements with the referee are shown to be unaffected by the team's ability. Instead, the match's present situation, such as unfavourable goal difference, makes disagreement with referees more likely to arise in the European soccer championship [12].

The study findings further revealed that the nature of the game has changed drastically over the years with many teams adopting aggressive plays [9]. This places a greater demand for referees during a match. It should be stressed here that the capacity to prioritize and analyse information at the proper moment in order to provide a suitable response to competing task demands reflects the dependability and perceptual-cognitive skill of the referees [13]. Nonetheless, the referees are shown to likely commit an unconscious bias due to the ever-increasing cognitive loads such as limited information, context, emotions as well as time pressures [14]. Unconscious bias in professional refereeing may imply that the acquired preconceptions that are rooted in their beliefs impact the way individual referees naturally deal with players and circumstances. It is inferred from the previous authors that judgement biases could be referred to as a subconscious state of mind overweighing some parts of information and underweighting or ignoring other information in relation to rules and circumstances [15]. Referees, for example, have been demonstrated to be influenced by audience emotions [16, 17].

2.5 Summary

The present investigation has identified the key fouls and misconducts-related activities in the European championship tournaments. A set of misconducts and fouls-related actions, namely fouls, challenges, challenges per foul, yellow cards, fouls per card, air challenges, ground challenges, dangerous play, misconduct, and attack wrecking, are demonstrated to be highly prevalent in the league. In addition, it was demonstrated that the nature of the game has changed with many clubs adopting aggressive plays, sabotage as well as misconduct-related offences. Moreover, it has been shown that referees are overloaded with consistent cognitive loads such as limited information, context, emotions as well as time pressures which often led to unconscious bias during judgements. It is, therefore, recommended that the training of referees should include aspects related to the heightened sabotage, misconduct, and fouls-related offences. This should be considered when naturing young referees to empower them to cope better during refereeing tasks.

References

1. M. Methenitis, Laws of the game. Video Game Policy **1**, 11–26 (2019). https://doi.org/10.4324/9781315748825-2
2. IFAB, *Statutes of the International Football Association Board* (2021)
3. R. Muazu Musa, A.P.P. Abdul Majeed, M.R. Abdullah, A.F. Ab. Nasir, M.H. Arif Hassan, M.A. Mohd Razman, Technical and tactical performance indicators discriminating winning and losing team in elite Asian beach soccer tournament. PLoS ONE **14**, e0219138 (2019). https://doi.org/10.1371/journal.pone.0219138
4. M.R. Abdullah, V. Eswaramoorthi, R.M. Musa, A.B.H.M. Maliki, N.A. Kosni, M. Haque, The effectiveness of aerobic exercises at difference intensities of managing blood pressure in essential hypertensive information technology officers. J. Young Pharm. **8**, 483–486 (2016). https://doi.org/10.5530/jyp.2016.4.27
5. M.A. Gipit, M.R.A. Charles, R.M. Musa, N.A. Kosni, A.B.H.M. Maliki, The effectiveness of traditional games intervention programme in the improvement of form one school-age children's motor skills related performance components. Mov. Health Exerc. **6**, 157–169 (2017)
6. Z. Taha, M. Haque, R.M. Musa, M.R. Abdullah, A. Maliki, N. Alias, N.A. Kosni, Intelligent prediction of suitable physical characteristics toward archery performance using multivariate techniques. J. Glob. Pharma Technol. **9**, 44–52 (2009)
7. M.R. Abdullah, A.B.H.M. Maliki, R.M. Musa, N.A. Kosni, H. Juahir, S.B. Mohamed, Identification and comparative analysis of essential performance indicators in two levels of soccer expertise. Int. J. Adv. Sci. Eng. Inf. Technol. **7**, 305–314 (2017). https://doi.org/10.18517/ijaseit.7.1.1150
8. V. Eswaramoorthi, M.R. Abdullah, R.M. Musa, A.B.H.M. Maliki, N.A. Kosni, N.B. Raj, N. Alias, H. Azahari, S.M. Mat-Rashid, H. Juahir, A multivariate analysis of cardiopulmonary parameters in archery performance. Hum. Mov. **19**, 35–41 (2018). https://doi.org/10.5114/hm.2018.77322
9. M.R. Abdullah, R.M. Musa, N.A. Kosni, A. Maliki, M. Haque, Profiling and distinction of specific skills related performance and fitness level between senior and junior Malaysian youth soccer players. Int. J. Pharm. Res. **8**, 64–71 (2016)

10. M.R. Razali, N. Alias, A. Maliki, R.M. Musa, L.A. Kosni, H. Juahir, Unsupervised pattern recognition of physical fitness related performance parameters among Terengganu youth female field hockey players. Int. J. Adv. Sci. Eng. Inf. Technol. **7**, 100–105 (2017)
11. R.T. Jewell, Estimating demand for aggressive play: the case of English premier league football. Int. J. Sport Finance **4**, 192–210 (2009)
12. K. Kempa, H. Rusch, *Misconduct and Leader Behaviour in Contests: New Evidence from European Football* (Philipps-University Marburg, School of Business and Economics, Marburg, 2016)
13. L.J. Moore, D.J. Harris, B.T. Sharpe, S.J. Vine, M.R. Wilson, Perceptual-cognitive expertise when refereeing the scrum in rugby union. J. Sports Sci. **37**, 1778–1786 (2019)
14. K.A. O'Brien, J. Mangan, The issue of unconscious bias in referee decisions in the national rugby league. Front. Sports Act. Living 260 (2021)
15. R. Moonesinghe, M.J. Khoury, A.C.J.W. Janssens, Most published research findings are false—but a little replication goes a long way. PLoS Med. **4**, e28 (2007)
16. K. Page, L. Page, Alone against the crowd: individual differences in referees' ability to cope under pressure. J. Econ. Psychol. **31**, 192–199 (2010)
17. M.K. Erikstad, B.T. Johansen, Referee bias in professional football: favoritism toward successful teams in potential penalty situations. Front. Sports Act. Living **2**, 19 (2020)

Chapter 3
Tactical and Misconduct Actions Leading to VAR Interventions in Top-Flights European Soccer Leagues

3.1 Overview

Referees are chosen to provide optimal performance in the sporting domain by maintaining a high level of integrity and objectivity. The referees are also expected to give leadership and guidance, apply common sense, analyse breaches, and evaluate regulations without any form of impartiality [1, 2]. The referees' capacity to prioritize and process information at the ideal time in order to provide the best judgement from a competing task demands reflects their reliability and perceptual-cognitive skill [3].

The referee has considerable authority in implementing the laws, including determining whether conduct falls under the broad categories of cautionable offences. As a result, judgments made by referees might be controversial. Some laws may specify the circumstances under which a warning should or must be issued, and various referee guidelines provide additional guidance. Nonetheless, referees are expected to use their common sense in making decisions [4]. For this and many other reasons, soccer referees are frequently accused of making incorrect decisions, particularly when it comes to delivering warnings or dismissals of players. Such incorrect decisions could emanate from bias that is characterized as systematic judgments in favour of particular teams, such as the home team or "big" clubs, that are seen as unjust [5]. This preferential treatment, which can occur because of purposeful prejudice, human mistake, or incompetence, could have a significant economic impact on the teams.

It is possible to reduce referee bias by decreasing the referees' control of the game and minimizing the number of "uncertain" calls that the referee often makes during a game. For instance, when a referee is unsure about which choice to make, he or she is more susceptible to being swayed by the crowd or the media. According to Pierluigi Collina, a former UEFA chief refereeing officer, adding additional assistant referees may result in better game control, mainly because they enable the other four (assistant) referees to concentrate on the main tasks [6]. In response to the calls by many stakeholders as well as the need for FIFA to preserve the nature of the game and reduce bias, video assistant referee (VAR) was introduced to support on-field referees in making decisions. In this paper, we examine the tactical and misconduct actions leading to VAR interventions in the European championship tournaments.

In this study, a dataset including 6232 matches from five consecutive seasons in the five best European leagues, namely the English Premier League, Spanish Laliga, Italian Serie A, French Ligue1, and German Bundesliga, was used. These datasets comprised both VAR and on-field referee functions such as issuing cautions in the form of yellow cards as well as dismissals of players via red cards based on different behaviours relating to misconducts and fouls. Before the commencement of the analysis, the data was preprocessed and checked for any missing information; rows that contained missing information are removed from the dataset [7, 8].

3.2 Clustering

A k-means clustering method is a form of cluster analysis approach that divides a set of data into k-predefined and non-overlapping subgroups called clusters, with only one group given to each data point [9–11]. The method seeks to make the inter-cluster data points as connected as feasible while maintaining the intra-cluster data points as different as possible. The cluster analysis was used to assign the VAR interventions into groups based on the number of occurrences during the matches. Euclidean distance was used as a distance metric to assign the formation of the two detected clusters, namely high video assistant referee intervention (H-VAR) and minimum assistant referee intervention (M-VAR).

3.3 Feature Selection

The present study employs many feature selection techniques, including information gain (IG), gain ratio, ReliefF, and chi-square. These techniques are used to extract the most crucial tactical and misconduct-related actions leading to the intervention of VAR. It is worth mentioning that such a strategy lowers overfitting, increases accuracy, and shortens training time by removing noise and redundant features [12–14]. The tactical and misconduct-related actions which consist of attack wrecking, ground challenges, misconduct, dangerous play, air challenges, challenges off the ball, handball, professional fouls as well as simulation/diving are analysed.

3.4 Machine Learning-Based Classification Model

In this investigation, the k-nearest neighbours (kNN) model is utilized to investigate its capability in classifying the type of VAR assistance. The dataset was divided based on a 70:30 split ratio that corresponds to training and testing, respectively. One hyperparameter of the kNN model was optimized, namely the number of neighbours, k by means of computing the mean accuracy on the training dataset via fivefold cross-validation technique. The value k was varied between one and 30, while the distance metric used was the Euclidean distance. It is worth noting that the performance of the optimized model was evaluated via the classification accuracy (CA), confusion matrix as well as the geometric mean (G-mean) score [15]. The evaluation of the model was carried out through Spyder v3.6.6, a Python (v3.7) IDE along with its related scikit-learn libraries.

3.5 Results and Discussion

Figure 3.1 depicts the classes identified through the k-means analysis with respect to the VAR interventions based on the tactical and misconduct-related offences committed by the players. It could be seen from the figure that a clear partition was established between the H-VAR and M-VAR. In other words, the mean occurrences of interventions provided by the VAR are substantially higher in H-VAR as compared to the M-VAR.

Table 3.1 gives the feature extractions technique employed to identify the essential tactical and misconduct-related offences that attract the attention of VAR. It could be

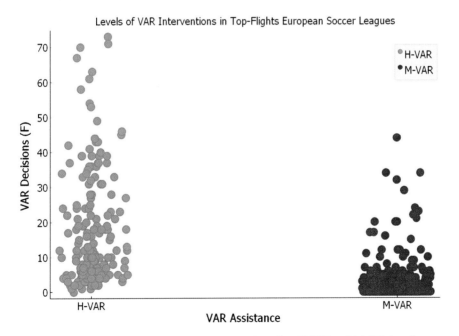

Fig. 3.1 Grouping of VAR assistance via k-means clustering. H-VAR = high VRA assistance, M-VAR = minimum VAR assistance

seen from the table that the most significant features identified from the IG features selection technique are attack wrecking, ground challenges, misconduct, dangerous play, air challenges, and challenges off the ball.

It could be seen from Fig. 3.2 that based on the hyperparameter investigated, i.e. the optimum value of k, three could yield a CA of 0.955 ($\pm$0.028) through the cross-validation evaluation. Upon employing the optimum number of k, the optimized kNN model could yield an accuracy of 92.3% on the test dataset. This observation suggests that the features selected are apt in providing a reasonably good classification of the VAR assistance. The confusion matrix of the test dataset illustrated in Fig. 3.3 demonstrates that only seven were misclassified for each VAR class. In addition, the G-mean score of 0.9123 indicates that a well-balanced accuracy is achieved.

It is demonstrated from the findings of the current investigation that actions which encapsulate sabotage and misconduct highly conjure VAR interventions. In other

Table 3.1 Feature selection

Tactical and misconducts actions	Infor. gain	Gain ratio	ReliefF	χ^2
Attack wrecking	**0.596**	0.298	0.167	318.561
Ground challenges	**0.521**	0.261	0.120	280.453
Misconduct	**0.512**	0.257	0.093	298.636
Dangerous play	**0.488**	0.244	0.106	242.642
Air challenges	**0.357**	0.204	0.056	370.431
Challenges off the ball	**0.207**	0.139	0.075	230.642
Handball	0.175	0.123	0.045	170.924
Professional foul	0.155	0.074	0.031	159.797
Simulation/diving	0.070	0.074	0.017	75.798

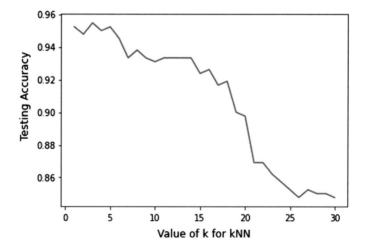

Fig. 3.2 Identifying the optimum number of k

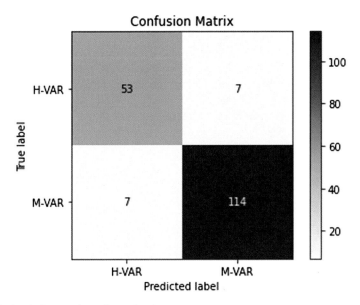

Fig. 3.3 Confusion matrix on the testing dataset

words, on-field referees are most unlikely to spot these actions without the intervention of VAR or might be difficult to issue appropriate judgement without the help of VAR. Misconduct may involve illegal behaviours which could occur at any moment, even when the ball is out of play, at half-time, and before and after the game [16]. The game of soccer has transformed into a high-pace game with many clubs adopting aggressive plays, sabotage as well as misconduct-related offences [17–19]. This places a high demand on the on-field referees to focus on several aspects of the game which might lead to unconscious errors in judgement [20]. On the other hand, FIFA declared a toughened rule on dangerous tackles by demanding that a deliberate tackle made from behind an opponent should be punished by a red card; hence for an on-field referee to ensure that the issue of the red card is based on a thorough judgement, there is a need for the referee to further consult VAR [21]. Consequently, it is not surprising that actions that translate to sabotage and aggressive tackles (attack wrecking, ground challenges, misconduct, dangerous play, air challenges, and challenges off the ball) are found to top the list of those actions that are most intervened by VAR.

3.6 Summary

The present investigation has established the tactical and misconduct-related offences that highly induced the interventions of VAR. These sets of actions that included attack wrecking, ground challenges, misconduct, dangerous play, air challenges, and challenges off the ball are found to be non-trivial in predicting the likelihood of on-field referees to consult VAR. It has also been demonstrated from the current finding that the optimized kNN model is able to yield an excellent prediction of the level of VAR assistance with respect to investigated parameters. As a result, it is advised that referee training should include features relating to high-level sabotage, misbehaviour, and fouls. This should be considered while naturing young referees in order to better equip them to cope with high challenging refereeing tasks.

References

1. C.D. Lirgg, D.L. Feltz, M.D. Merrie, Self-efficacy of sports officials: a critical review of the literature. J. Sport Behav. **39** (2016)
2. D.J. Hancock, L.J. Martin, M.B. Evans, K.F. Paradis, Exploring perceptions of group processes in ice hockey officiating. J. Appl. Sport Psychol. **30**, 222–240 (2018)
3. L.J. Moore, D.J. Harris, B.T. Sharpe, S.J. Vine, M.R. Wilson, Perceptual-cognitive expertise when refereeing the scrum in rugby union. J. Sports Sci. **37**, 1778–1786 (2019)
4. M. Methenitis, Laws of the game. Video Game Policy **1**, 11–26 (2019). https://doi.org/10.4324/9781315748825-2
5. T. Dohmen, J. Sauermann, Referee bias. J. Econ. Surv. **30**, 679–695 (2016)
6. A. Albanese, S. Baert, O. Verstraeten, Twelve eyes see more than eight. Referee bias and the introduction of additional assistant referees in soccer. PLoS ONE **15**, e0227758 (2020)
7. M.R. Abdullah, V. Eswaramoorthi, R.M. Musa, A.B.H.M. Maliki, N.A. Kosni, M. Haque, The effectiveness of aerobic exercises at difference intensities of managing blood pressure in essential hypertensive information technology officers. J. Young Pharm. **8**, 483–486 (2016). https://doi.org/10.5530/jyp.2016.4.27
8. M.A. Gipit, M.R.A. Charles, R.M. Musa, N.A. Kosni, A.B.H.M. Maliki, The effectiveness of traditional games intervention programme in the improvement of form one school-age children's motor skills related performance components. Mov. Health Exerc. **6**, 157–169 (2017)
9. H. Azahari, H. Juahir, M.R. Abdullah, R.M. Musa, V. Eswaramoorthi, N. Alias, S.M. Mat-Rashid, N.A. Kosni, A.B.H.M. Maliki, N.B. Raj, A multivariate analysis of cardiopulmonary parameters in archery performance. Hum. Mov. **19**, 35–41 (2019). https://doi.org/10.5114/hm.2018.77322
10. M.R. Razali, N. Alias, A. Maliki, R.M. Musa, L.A. Kosni, H. Juahir, Unsupervised pattern recognition of physical fitness related performance parameters among Terengganu youth female field hockey players. Int. J. Adv. Sci. Eng. Inf. Technol. **7**, 100–105 (2017)
11. Z. Taha, M. Haque, R.M. Musa, M.R. Abdullah, A. Maliki, N. Alias, N.A. Kosni, Intelligent prediction of suitable physical characteristics toward archery performance using multivariate techniques. J. Glob. Pharma Technol. **9**, 44–52 (2009)
12. R.M. Musa, A.P.P. Abdul Majeed, A. Musa, M.R. Abdullah, N.A. Kosni, M.A.M. Razman, An information gain and hierarchical agglomerative clustering analysis in identifying key performance parameters in elite beach soccer (2021). https://doi.org/10.1007/978-981-15-7309-5_26
13. R. Muazu Musa, A.P.P. Abdul Majeed, N.A. Kosni, M.R. Abdullah, Key performance indicators in elite beach soccer, in *Springer Briefs in Applied Sciences and Technology* (2020), pp. 13–19. https://doi.org/10.1007/978-981-15-3219-1_2
14. R.M. Musa, A.P.P.A. Majeed, N.A. Kosni, M.R. Abdullah, *Machine Learning in Team Sports: Performance Analysis and Talent Identification in Beach Soccer & Sepak-Takraw* (Springer Nature, 2020)
15. A. Kulkarni, D. Chong, F.A. Batarseh, Foundations of data imbalance and solutions for a data democracy, in *Data Democracy* (Elsevier, 2020), pp. 83–106

16. R.T. Jewell, Estimating demand for aggressive play: the case of English Premier League football. Int. J. Sport Finance **4**, 192–210 (2009)
17. M.R. Abdullah, R.M. Musa, N.A. Kosni, A. Maliki, M. Haque, Profiling and distinction of specific skills related performance and fitness level between senior and junior Malaysian youth soccer players. Int. J. Pharm. Res. **8**, 64–71 (2016)
18. M.R. Abdullah, A. Maliki, R.M. Musa, N.A. Kosni, H. Juahir, S.B. Mohamed, Identification and comparative analysis of essential performance indicators in two levels of soccer expertise. Int. J. Adv. Sci. Eng. Inf. Technol. **7**, 305–314 (2017)
19. R. Muazu Musa, A.P.P. Abdul Majeed, M.R. Abdullah, A.F. Ab. Nasir, M.H. Arif Hassan, M.A. Mohd Razman, Technical and tactical performance indicators discriminating winning and losing team in elite Asian beach soccer tournament. PLoS ONE **14**, e0219138 (2019)
20. K.A. O'Brien, J. Mangan, The issue of unconscious bias in referee decisions in the national rugby league. Front. Sports Act. Living **260** (2021)
21. IFAB, *Statutes of the International Football Association Board* (2021)

Chapter 4
Positional Events Incidences Leading to VAR Intervention in European Soccer Leagues Games

4.1 Overview

Following the successful roll-out of Goal Line Technology during the 2012 FIFA World Cup, the usage of technology in soccer has grown rapidly. Global positioning system (GPS) technology has been employed in professional sports, both during training and competition. Teams may follow players' movements on the field and gather large quantities of data on their performance. Such technology generally known as "Electronic Performance and Tracking System (EPTS)" devices can quantify and accumulate data in relation to players' running speed, distance ran, location on the field, heart rate, and body's work rate [1].

The International Football Association Board (IFAB) launched the video assistant referee (VAR) system as football's first use of video technology to make better judgements. At the 132nd Annual General Meeting in Zurich on 3 March 2018, the IFAB, the organization that governs the Laws of the Game of association football, fully authorized the use of VAR. This was done to improve the sport's fairness and integrity [2].

GPS data might be used to track a player's position on the field, assisting in identifying the most common territories covered and providing insights into how well particular areas were utilized. This data may then be used to tailor the training and development of specific models [3]. The type of data acquired by GPS trackers might vary greatly depending on the supplier and the needs of the team using the data. As with other aspects of performance analysis, GPS data must be used and analysed effectively in the context of the sport, player, or scenario. Consequently, we intend to utilize the data provided by the GPS to extract information related to positional events occurrences leading to VAR intervention in the European championship league. Essentially, we sought to predict the interventions of VAR at various locations of the pitch where actions occurred.

4.2 Events Variables

In this study, a dataset comprising 6232 matches from five consecutive seasons from the five best European leagues, namely the English Premier League, Spanish Laliga, Italian Serie A, French Ligue1, and German Bundesliga, was used. These datasets comprised both VAR and on-field referee activities at various locations on the pitch. Positional event data of the referees and the VAR interventions consist of the average distance to the event, distance to fouls, distance to fouls leading to cards distance to goals, avg. referee distance to penalty box + 2 m around fouls, decisions in key zones, distance to 8 m, decisions in key zones, distance to 8–20 m, decisions in key zones, distance more than 2 m, average referee distance in other zones, decisions in other zones from close range (less than 25 m), and decisions in other zones from a distance (more than 25 m) were used to ascertain the likelihood of VAR interventions concerning actions occurrences at specific positions. It is worth mentioning that before the commencement of the analysis, the data was preprocessed and checked for any missing information; rows that contained missing information are removed from the dataset [4–6].

4.3 Machine Learning-Based Classification Analysis

In this particular investigation, the support vector machine (SVM) model is employed to investigate its ability in demarcating the type of VAR intervention. The 70:30 split ratio was utilized for training and testing, respectively. The hyperparameters of the SVM model were not optimized in this study to gauge its natural ability in demarcating the classes. The classification accuracy (CA), confusion matrix, and the geometric mean (G-mean) score [7] were used to evaluate the performance of the developed model. A Python (3.7) IDE, namely Spyder v3.6.6, was used to carry out the analysis of the present study along with its typical scikit-learn libraries.

4.4 Results and Discussion

The difference in the assistance rendered by the VAR based on event occurrences during match play is shown in Fig. 4.1. It was demonstrated from Fig. 4.1a that the average referee's distance to the penalty box plus 2 m around fouls attracted more VAR assistance compared to the average referee's distance in other zones except penalty box plus 2 m around. Similarly, in Fig. 4.1b, referees' decisions in key zones, distance 8–20 m, evoke more VAR assistance in comparison with the referees' key zones, distance to 8 m. Moreover, in Fig. 4.1c, referees' decisions in other zones from a close range of less than 25 m have recorded the highest VAR assistance when compared to the decision in key zones distance to 8 m and distances more than 20 m.

Overall, it could be deduced from the figures that VAR assistance is likely to be sought after when events occur in key zones. Additionally, on-field referees have a higher tendency of consulting VAR when they are relatively far away from where an action occurs.

It was shown from the study that based on the parameters selected, the vanilla SVM model is able to yield a CA of 88.12% and 83.43% for training and testing, respectively. Although slight overfitting is observed, nonetheless the G-mean score (0.8458) indicates that a well-balanced dataset accuracy is achieved in classifying the M-VAR and H-VAR. Figure 4.2a, b illustrates the confusion matrix for both the training and testing datasets.

Since sports contain moments when referees and officials make mistakes, the use of technology has become increasingly important in recent years. Recent technological advances in high-performance sports have helped to reduce some of these errors [8]. Throughout the last few years, it has become apparent that the soccer game is evolving into a fast-moving sport due to the increased level of fitness among players and the evolution of tactics in recent years, which has placed more pressure on the referees to be at their best since their decisions are scrutinized owing to the wider recognition of the sport globally [4, 5, 9–12].

Due to the implementation of VAR, soccer has advanced one step closer to perfection. The VAR is frequently used to overturn game-changing judgments, such as offside rulings, but it may also be used to assess whether a goal is erroneously disallowed, or an infraction happened before the goal was scored [13]. Close offside calls are the most typical reason for VAR consultation after a goal has been scored; however, shirt-pulling and other infringements can also be factors in goals being disallowed. VAR is also found to be useful during penalty checking, a straight red card, or a case of mistaken identification [14]. This aids in evaluating refereeing judgments that were not witnessed in real time.

4.5 Summary

The current study has identified the positional event occurrences that lead to the assistance of VAR. It was demonstrated from the findings of the study that the average referee's distance to the penalty box plus 2 m around fouls attracted more VAR assistance compared to the average referee's distance in other zones except penalty box plus 2 m around. Similarly, referees' decisions in key zones, distance 8–20 m, evoke more VAR assistance in comparison with the referees' key zones, distance to 8 m. Moreover, referees' decisions in other zones from a close range of less than 25 m are shown to invite higher VAR interventions when compared to the decision in key zones distance to 8 m and distances more than 20 m. Additionally, the findings further demonstrated that the SVM-based model is able to yield a good prediction of the VAR intervention based on the zonal positions examined in the study.

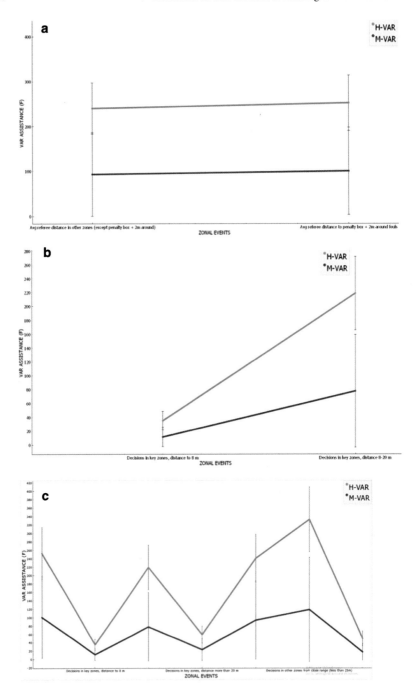

Fig. 4.1 Differences in VAR assistance between zones

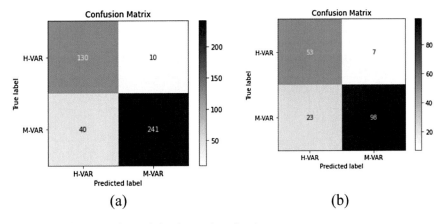

Fig. 4.2 Confusion matrix **a** training dataset, **b** testing dataset

References

1. C. Carling, J. Bloomfield, L. Nelsen, T. Reilly, The role of motion analysis in elite soccer. Sports Med. **38**, 839–862 (2008)
2. IFAB, *Statutes of the International Football Association Board* (2021)
3. A. Mendez-Villanueva, M. Buchheit, B. Simpson, P.C. Bourdon, Match play intensity distribution in youth soccer. Int. J. Sports Med. **34**, 101–110 (2013)
4. H. Azahari, H. Juahir, M.R. Abdullah, R.M. Musa, V. Eswaramoorthi, N. Alias, S.M. Mat-Rashid, N.A. Kosni, A.B.H.M. Maliki, N.B. Raj, A multivariate analysis of cardiopulmonary parameters in archery performance. Hum. Mov. **19**, 35–41 (2019). https://doi.org/10.5114/hm.2018.77322
5. Z. Taha, M. Haque, R.M. Musa, M.R. Abdullah, A. Maliki, N. Alias, N.A. Kosni, Intelligent prediction of suitable physical characteristics toward archery performance using multivariate techniques. J. Glob. Pharma Technol. **9**, 44–52 (2009)
6. M.A. Gipit, M.R.A. Charles, R.M. Musa, N.A. Kosni, A.B.H.M. Maliki, The effectiveness of traditional games intervention programme in the improvement of form one school-age children's motor skills related performance components. Mov. Health Exerc. **6**, 157–169 (2017)
7. A. Kulkarni, D. Chong, F.A. Batarseh, Foundations of data imbalance and solutions for a data democracy, in *Data Democracy* (Elsevier, 2020), pp. 83–106
8. D.K. Ntege, Impact of the use of technology based referee system in football: a case of Spanish La Liga video assistant referee (2020)
9. M.R. Abdullah, R.M. Musa, N.A. Kosni, A. Maliki, M. Haque, Profiling and distinction of specific skills related performance and fitness level between senior and junior Malaysian youth soccer players. Int. J. Pharm. Res. **8**, 64–71 (2016)
10. R. Muazu Musa, A.P.P. Abdul Majeed, M.R. Abdullah, A.F. Ab. Nasir, M.H. Arif Hassan, M.A. Mohd Razman, Technical and tactical performance indicators discriminating winning and losing team in elite Asian beach soccer tournament. PLoS ONE **14**, e0219138 (2019)
11. M.R. Razali, N. Alias, A. Maliki, R.M. Musa, L.A. Kosni, H. Juahir, Unsupervised pattern recognition of physical fitness related performance parameters among Terengganu youth female field hockey players. Int. J. Adv. Sci. Eng. Inf. Technol. **7**, 100–105 (2017)

12. M.R. Abdullah, A. Maliki, R.M. Musa, N.A. Kosni, H. Juahir, S.B. Mohamed, Identification and comparative analysis of essential performance indicators in two levels of soccer expertise. Int. J. Adv. Sci. Eng. Inf. Technol. **7**, 305–314 (2017)
13. M. Methenitis, Laws of the game. Video Game Policy **1**, 11–26 (2019). https://doi.org/10.4324/9781315748825-2
14. K. Page, L. Page, Alone against the crowd: individual differences in referees' ability to cope under pressure. J. Econ. Psychol. **31**, 192–199 (2010)

Chapter 5
Decisions Error of Top European Leagues Soccer Leagues Referees at Specific Time of Match Play

5.1 Overview

Referees have one of the most difficult jobs in international football, since every judgement they make may have a significant impact on the outcome of the game, as well as the entire season. Every referee must make difficult judgments and will be criticized at some point by players, managers, and supporters. The most successful referees are the ones who can handle criticism and deliver the right decision most of the time. It is worth noting that referee instincts are critical to their decision-making; they may either lead to accurate decisions or raise the likelihood of making severe errors during a game. The referee's decisions have a significant influence on the game, and the referees need to apply common sense on the field to aid in decision-making [1]. However, there have been several cases in which referees are caught up committing serious errors during decision-making [2].

Decision-making is influenced by social factors, and decision-makers are often reported to have their distinctive interpretations, attitudes, hidden biases, forecasts, and shown preferences [3–5]. Surprisingly, illusionary memories have been proven to influence decision-making during gameplay [6]. A recent study has also found that biases or shortcut tactics used to handle complicated information can occur as a result of decision-makers attempting to swiftly recognize and process fast-paced visual, linguistic, and behavioural signals [7, 8]. Since referees are necessitated to make decisions under time restrictions, in complicated situations, and sometimes under confusing conditions, a tendency or prejudice for or against an individual player or team appears to be very likely [9].

The variations in decision-making by referees demonstrated some elements of inconsistencies in the delivery of judgement as well as enforcement of certain punishments both among the elite and highly experienced professionals [10]. Although these variations might not necessarily occur as a consequence of unconscious bias, some specific situations may then explain what triggered referees in making judgement during match play. For instance, the presents of crowds, home team advantage, and referees' prejudice have been shown to have an impact on referees' decisions.

R. Muazu Musa et al., *Data Mining and Machine Learning in High-Performance Sport*, SpringerBriefs in Applied Sciences and Technology, https://doi.org/10.1007/978-981-19-7049-8_5

However, there is a limited investigation of the referee's decision at a specific time of play. Hence, in this investigation, we explored the occurrences of decision errors with reference to specific time play during the game in European championship tournaments.

5.2   Clustering

The Louvain cluster analysis was utilized in this investigation. The number of decision errors committed by the referees throughout five seasons was initially clustered to obtain the referees' levels of decision errors. Subsequently, the errors committed at a specific time of play, namely first half injury time, second half injury time, injury time, and ball in play time, were used to ascertain the differences of the initially developed clusters in the number of errors committed at varying time play [11–13]. The clusters were treated as dependent variables, while the decision errors committed by the referees at different times were considered as independent variables [14–17].

5.3   Classification

In this exploratory study, a rudimentary classifier, i.e. logistic regression (LR) is used to investigate its efficacy in discerning decision errors. A split ratio of 70:30 was employed on the dataset for training and testing, respectively. The hyperparameters of the LR model were set as default for the present study. The performance of the model is illustrated via the classification accuracy (CA), confusion matrix as well as the geometric mean (G-mean) score [18]. The analysis was carried out by utilizing a Python (3.7) IDE, viz. Spyder v 3.6.6 along with associated scikit-learn libraries.

5.4   Results and Discussion

Figure 5.1 projects the level of decision error obtained via the Louvain clustering analysis. It could be detected that two clusters were determined based on the similarities in the characteristics of the parameters evaluated. The figures further demonstrated a transparent partition of the clusters which is essential for the assignation of the identity to each of the clusters, viz. high-decision error (HDE) as well as low-decision error (LDE), respectively.

Figure 5.2 depicts the mean performance differences plots of the assigned clusters with respect to a specific time during match play. It could be observed from line plots that the occurrences of decision errors are highly skewed concerning specific time play during the game. High-decisions error is found to be substantially higher during ball in play time followed by injury time. Although there is no marked difference in

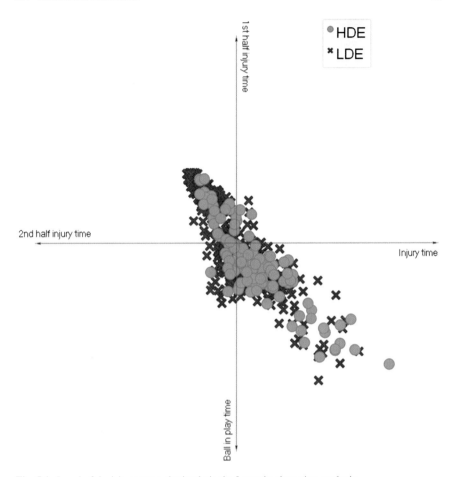

Fig. 5.1 Level of decisions error obtained via the Louvain clustering analysis

decisions error at both first and second half injury time, nonetheless, a slight increase in decisions error is observed in the course of second half injury time. Overall, it could be postulated from the figure that high occurrences of decision error are more likely during a ball in play time, injury time, and possibly at the second half injury time as compared to first half injury time play.

The findings from the present investigation demonstrated that there is a strong association between decision errors and specific time of match play. To ensure a fair outcome, referees have several ways of imparting justice: by validating goals, awarding penalties, sending off players, making foul calls, and determining injury time. The actions of a referee can significantly alter the outcome of a game. Referees may make a huge difference in the outcome of a game by not awarding goals, letting red cards stand, and not calling off-sides that result in goals. It is observed from the findings of this investigation that referees are likely to make such errors during

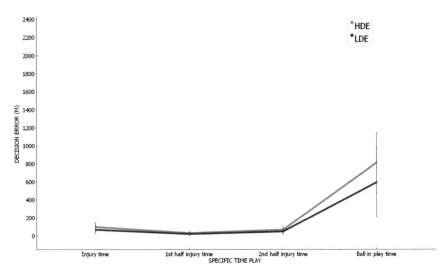

Fig. 5.2 Differences in decisions error across varying time play in European championship tournaments

gameplay time. Since during the ball in play time, both teams are not constrained by period and the desire to push for more goals or chances might be lesser; hence, referees may be inclined to commit more errors as the likelihood for such errors to go unnoticed is high. On the other hand, it is shown that referees make higher errors in injury time. This may be associated with many factors that could be either based on bias, human errors, or simply a result of the presence of crowds [19].

Moreover, to further predict the occurrences of decision errors with respect to the specific time play, it was demonstrated that the LR model could yield a CA of 79.57% and 77.35% for training and testing, respectively. Although a relatively good CA was obtained, nonetheless, upon scrutinizing the confusion matrix as depicted in Fig. 5.3a, b, it is apparent that a high misclassification transpired in identifying HDE. This observation is supported by the G-mean score that is recorded at 0.2232, suggesting the skewness of the prediction. This may be contributed to the imbalanced nature of the data.

5.5   Summary

It has been revealed from the findings of the current investigation that there is a correlation between decision errors and specific time in match play. It is demonstrated that high-decisions error is mostly committed by the referees during ball in play time, injury time, and possibly at the second half injury time as compared to first half injury time play. Moreover, the LR-based classification model was found to be effective in the identification of high-decision errors at a specific time to play. While it is

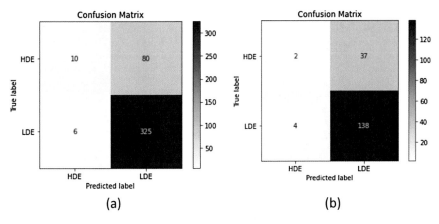

Fig. 5.3 Confusion matrix **a** training dataset, **b** testing dataset

noted that refereeing especially at the elite level is not an easy task and the referees as a human might likely make mistakes, however, the findings of this investigation could be used to educate referees on the challenges of refereeing during a specific time in elite level matches which is non-trivial in mapping out strategies to curtail its occurrences.

References

1. IFAB, *Statutes of the International Football Association Board* (2021)
2. T. Dohmen, J. Sauermann, Referee bias. J. Econ. Surv. **30**, 679–695 (2016)
3. M. Thuraisingham, *The Secret Life of Decisions: How Unconscious Bias Subverts Your Judgement* (Routledge, 2017)
4. M.G. Rhodes, A.E. Witherby, A.D. Castel, K. Murayama, Explaining the forgetting bias effect on value judgments: the influence of memory for a past test. Mem. Cognit. **45**, 362–374 (2017)
5. P. Agarwal, *Sway: Unravelling Unconscious Bias* (Bloomsbury Publishing, 2021)
6. H. Hill, S. Windmann, Examining event-related potential (ERP) correlates of decision bias in recognition memory judgments. PLoS ONE **9**, e106411 (2014)
7. A. Albanese, S. Baert, O. Verstraeten, Twelve eyes see more than eight. Referee bias and the introduction of additional assistant referees in soccer. PLoS ONE **15**, e0227758 (2020)
8. L.J. Moore, D.J. Harris, B.T. Sharpe, S.J. Vine, M.R. Wilson, Perceptual-cognitive expertise when refereeing the scrum in rugby union. J. Sports Sci. **37**, 1778–1786 (2019)
9. M.K. Erikstad, B.T. Johansen, Referee bias in professional football: favoritism toward successful teams in potential penalty situations. Front. Sports Act. Living **2**, 19 (2020)
10. K.A. O'Brien, J. Mangan, The issue of unconscious bias in referee decisions in the national rugby league. Front. Sports Act. Living 260 (2021)
11. H. Azahari, H. Juahir, M.R. Abdullah, R.M. Musa, V. Eswaramoorthi, N. Alias, S.M. Mat-Rashid, N.A. Kosni, A.B.H.M. Maliki, N.B. Raj, A multivariate analysis of cardiopulmonary parameters in archery performance. Hum. Mov. **19**, 35–41 (2019). https://doi.org/10.5114/hm. 2018.77322

12. M.R. Razali, N. Alias, A. Maliki, R.M. Musa, L.A. Kosni, H. Juahir, Unsupervised pattern recognition of physical fitness related performance parameters among Terengganu youth female field hockey players. Int. J. Adv. Sci. Eng. Inf. Technol. **7**, 100–105 (2017)
13. Z. Taha, M. Haque, R.M. Musa, M.R. Abdullah, A. Maliki, N. Alias, N.A. Kosni, Intelligent prediction of suitable physical characteristics toward archery performance using multivariate techniques. J. Glob. Pharma Technol. **9**, 44–52 (2009)
14. R. Muazu Musa, A.P.P. Abdul Majeed, M.R. Abdullah, A.F. Ab. Nasir, M.H. Arif Hassan, M.A. Mohd Razman, Technical and tactical performance indicators discriminating winning and losing team in elite Asian beach soccer tournament. PLoS ONE **14**, e0219138 (2019)
15. M.R. Abdullah, A. Maliki, R.M. Musa, N.A. Kosni, H. Juahir, S.B. Mohamed, Identification and comparative analysis of essential performance indicators in two levels of soccer expertise. Int. J. Adv. Sci. Eng. Inf. Technol. **7**, 305–314 (2017)
16. M.R. Abdullah, R.M. Musa, N.A. Kosni, A. Maliki, M. Haque, Profiling and distinction of specific skills related performance and fitness level between senior and junior Malaysian youth soccer players. Int. J. Pharm. Res. **8**, 64–71 (2016)
17. M.A. Gipit, M.R.A. Charles, R.M. Musa, N.A. Kosni, A.B.H.M. Maliki, The effectiveness of traditional games intervention programme in the improvement of form one school-age children's motor skills related performance components. Mov. Health Exerc. **6**, 157–169 (2017)
18. A. Kulkarni, D. Chong, F.A. Batarseh, Foundations of data imbalance and solutions for a data democracy, in *Data Democracy* (Elsevier, 2020), pp. 83–106
19. K. Page, L. Page, Alone against the crowd: individual differences in referees' ability to cope under pressure. J. Econ. Psychol. **31**, 192–199 (2010)

Chapter 6
Prevalence and Differences of Decisions Error in Top-Class European Soccer Leagues

6.1 Overview

In association football, commonly known as soccer, referees play an enormous role in determining the outcome of a game, such as determining goals, penalties, red and yellow cards, offside calls, and fouls [1]. If a referee is prejudiced or often commits errors in delivering judgement, his or her conduct on the field may be detrimental to the sport's best interests. As this prejudice and error grow, the sport may lose appeal as spectators may cast a doubt on the fairness of the competition, resulting in financial loss for the league [2]. It is difficult or rather impossible for the organizers to control referees. It is worth noting that the act of refereeing is not an easy task and referees may sometimes make mistakes as humans. However, tracking the performance of the referees' judgement is essential for the application of the appropriate intervention.

Performance evaluation is frequently used as a diagnostic instrument for evaluating the performance of teams or athletes [3–5]. In an age of increasing electronic surveillance, the errors made by the referees are often highlighted to provide the organizers to educate and train referees to deliver consistent and objective rulings. However, many issues such as bias and systemic errors could be hidden in very complex and strategic ways, and referees, even at the elite level, are rational agents who can learn how their organization functions and adjusts accordingly, which makes detection difficult [6, 7].

It is expected that the standard of refereeing is high across all the European leagues due to their popularity. Moreover, the presence of seven officials on the field during a typical soccer game should reflect relatively accurate decisions amidst every action as well as any specific infractions. However, it is often argued that some leagues have better officiating personnel than others in European tournaments. It is against this background that the current study is undertaken to explore the variations of decisions error across the leagues.

© The Author(s), under exclusive license to Springer Nature Singapore Pte Ltd. 2022
R. Muazu Musa et al., *Data Mining and Machine Learning in High-Performance Sport*,
SpringerBriefs in Applied Sciences and Technology,
https://doi.org/10.1007/978-981-19-7049-8_6

6.2   Types of Decisions Error Across the Leagues

In the present study, event variables that consist of foul not detected, yellow cards not shown, red cards not shown offside not detected as well as goals disallowed are selected for investigation. These variables are considered the types of decisions error that are vital for assessing the effectiveness of referees' decisions across the leagues. Red and yellow cards, offsides, foul calls, and a legitimate goal otherwise disallowed could change the course of a game and its outcome. These event variables portray instances where referees could exert their influence for changing the game's outcome. For instance, a slight error of a referee such as an unfair yellow card shown to a defender early in a game could restrict the player throughout the match thereby forcing the player to play with extra caution, which may increase the likelihood of a score from the opposing team. It is worth highlighting that before the commencement of the analysis, the data was preprocessed and checked for any missing information; rows that contained missing information are removed from the dataset [8–10]. In the data analysis stage, the leagues were treated as dependent variables, while the decision error variables were considered as independent variables [10–13].

6.3   Results and Discussion

Table 6.1 reveals the event variables that distinguish the referees in their decision-making ability. It could be seen that out of the four investigated indicators, three were found to be statistically significant between the leagues, viz. foul not detected, offside not detected, and goal disallowed $p < 0.05$. On the other hand, yellow cards not shown are observed to be not significantly different across the leagues $p > 0.05$. It could then be postulated that decisions error differs across the leagues.

Figure 6.1 portrays the mean differences plots of the assessed variables across the leagues. It could be observed from the boxplots that the Italian Serie A accumulated the highest number of fouls not detected in 192 event actions. The English Premier League recorded the highest in the number of offsides not detected and the second highest in the number of errors in fouls not detected as well as goals disallowed within a total of 96 instances. The Spanish Laliga referees recorded the highest number of goals disallowed in 100 instances. The German Bundesliga is observed to be the third highest in the frequency of fouls not detected. Remarkably, the French Ligue1 referees are found to record the lowest-decisions error with respect to the investigated variables. Overall, it could be posited that the English Premier League referees committed more errors in comparison with the referees of Bundesliga, Laliga as well as Ligue1.

The overall findings from this study demonstrated that decisions error made by the referees varied across leagues of the European championship. A combination of certain event parameters that constitute foul not detected, offsides not detected, yellow cards not shown, red cards not shown, and goals disallowed could potentially

Table 6.1 Decisions error differences across European leagues referees

Events	Bundesliga (123)		Laliga (100)		League1 (91)		Premier (96)		Series A (192)		P-value
	Mean	Std. dev.	Mean	Std. dev.	Mean	Std. dev.	Mean	Std. dev.	Mean	Std. dev.	
Foul not detected	0.398	0.744	0.340	0.755	0.165	0.563	0.510	0.871	0.531	1.013	0.008
Yellow cards (not shown)	0.033	0.178	0.060	0.278	0.066	0.389	0.031	0.175	0.047	0.276	0.846
Offside not detected	0.244	0.518	0.300	0.560	0.143	0.382	0.427	0.736	0.188	0.476	0.002
Goal disallowed	0.268	0.559	0.550	0.730	0.176	0.529	0.458	0.845	0.250	0.655	0.000

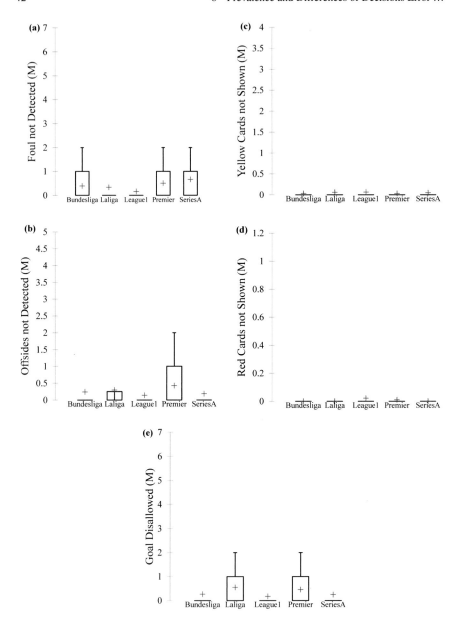

Fig. 6.1 Decisions error differences among top European leagues **a** foul not detected, **b** offsides not detected, **c** yellow cards not shown, **d** red cards not shown, and **e** goals disallowed

serve as a distinctive marker between the referees in their delivery of judgement during match play. Moreover, it is determined that English Premier League referees committed more errors as compared to referees in the Bundesliga, Laliga, Serie A, and League1. Interestingly, the referees from the French League1 display the lowest decision error relative to the variables examined. However, it should be noted that this finding is limited to the specific performance indicators evaluated; hence, the finding could not be generalized to the other parameters that are not included in this study.

6.4 Summary

The present investigation has established that certain event variables could distinguish the top European league referees in their decision-making ability. It was observed from the findings that three event variables that consist of foul not detected, offside not detected, and goal disallowed could significantly distinguish the referees in their ability to deliver appropriate decisions. Based on the findings of the current investigation, it is evident to surmise that the English Premier League referees committed more errors in comparison with the referees of Bundesliga, Laliga, Serie A as well as Ligue1.

References

1. M. Fontenla, G.M. Izón, The effects of referees on the final score in football. Estud. Econ. **35**, 79–98 (2018)
2. K. Page, L. Page, Alone against the crowd: individual differences in referees' ability to cope under pressure. J. Econ. Psychol. **31**, 192–199 (2010)
3. M.A. Gomez, C. Lago-Peñas, J. Viaño, I. González-Garcia, Effects of game location, team quality and final outcome on game-related statistics in professional handball close games. Kinesiol. Int. J. Fundam. Appl. Kinesiol. **46**, 249–257 (2014)
4. A. Arnason, S.B. Sigurdsson, A. Gudmundsson, I. Holme, L. Engebretsen, R. Bahr, Physical fitness, injuries, and team performance in soccer. Med. Sci. Sports Exerc. **36**, 278–285 (2004)
5. R. Muazu Musa, A.P.P. Abdul Majeed, M.R. Abdullah, A.F. Ab. Nasir, M.H. Arif Hassan, M.A. Mohd Razman, Technical and tactical performance indicators discriminating winning and losing team in elite Asian beach soccer tournament. PLoS ONE **14**, e0219138 (2019). https://doi.org/10.1371/journal.pone.0219138
6. P. Agarwal, *Sway: Unravelling Unconscious Bias* (Bloomsbury Publishing, 2021)
7. A. Osório, Performance evaluation: subjectivity, bias and judgment style in sport. Group Decis. Negot. **29**, 655–678 (2020)
8. V. Eswaramoorthi, M.R. Abdullah, R.M. Musa, A.B.H.M. Maliki, N.A. Kosni, N.B. Raj, N. Alias, H. Azahari, S.M. Mat-Rashid, H. Juahir, A multivariate analysis of cardiopulmonary parameters in archery performance. Hum. Mov. **19**, 35–41 (2018). https://doi.org/10.5114/hm.2018.77322
9. Z. Taha, M. Haque, R.M. Musa, M.R. Abdullah, A. Maliki, N. Alias, N.A. Kosni, Intelligent prediction of suitable physical characteristics toward archery performance using multivariate techniques. J. Glob. Pharma Technol. **9**, 44–52 (2009)

10. M.A. Gipit, M.R.A. Charles, R.M. Musa, N.A. Kosni, A.B.H.M. Maliki, The effectiveness of traditional games intervention programme in the improvement of form one school-age children's motor skills related performance components. Mov. Health Exerc. **6**, 157–169 (2017)
11. M.R. Abdullah, R.M. Musa, N.A. Kosni, A. Maliki, M. Haque, Profiling and distinction of specific skills related performance and fitness level between senior and junior Malaysian youth soccer players. Int. J. Pharm. Res. **8**, 64–71 (2016)
12. M.R. Abdullah, A.B.H.M. Maliki, R.M. Musa, N.A. Kosni, H. Juahir, S.B. Mohamed, Identification and comparative analysis of essential performance indicators in two levels of soccer expertise. Int. J. Adv. Sci. Eng. Inf. Technol. **7**, 305–314 (2017). https://doi.org/10.18517/ija seit.7.1.1150
13. M.R. Razali, N. Alias, A. Maliki, R.M. Musa, L.A. Kosni, H. Juahir, Unsupervised pattern recognition of physical fitness related performance parameters among Terengganu youth female field hockey players. Int. J. Adv. Sci. Eng. Inf. Technol. **7**, 100–105 (2017)

Chapter 7
Relationship Between Match Loading and Decisions Error Among Top-Class European Soccer Leagues Referees

7.1 Overview

A referee is an individual in charge of interpreting and implementing the Laws of the Game during a soccer match. The referee is the match official with the ability to start and stop play as well as execute disciplinary punishment on players and coaches during match play. Sports referees have a high level of responsibility due to their obligation to make the correct decision at any time during the match play. Extra match officials, i.e. two assistant referees, a fourth official, two additional assistant referees, a reserve assistant referee, a video assistant referee (VAR), and at least one assistant VAR (AVAR) may be chosen. They will help the referee regulate the game in line with the Laws of the Game; however, the referee would always reserve the right to deliver the final decision. It is worth noting that the referee, assistant referees, fourth official, additional assistant referees, and reserve assistant referees are categorized as the "on-field" match officials. The VAR and AVAR are considered and serve as "video" match officials (VMOs) who assist the referee in line with the Laws of the Game and the VAR procedure [1].

Referees play an important part in regulating game dynamics in today's sports. For the referees to be successful, extensive knowledge of the rules, as well as well-developed psychological and perceptual skills, coupled with optimal physical abilities are necessary. The physical and physiological prowess of the referee could allow them to move on the pitch throughout a match in order to position themselves in a timely and correct manner in each action and to have sufficient criteria to solve each play in the best way possible [2].

There has been a growing interest among sports scientists in studying referees' behaviour in recent years. This has culminated in more research on quantifying and describing the external and internal loads of the referees during match play [3]. It has been reported during match play that the referees are subjected to multiple intermittent movements throughout the play, spending time standing, walking, sprinting at various intensities, and moving in unconventional ways [4, 5]. However, taking

R. Muazu Musa et al., *Data Mining and Machine Learning in High-Performance Sport*, SpringerBriefs in Applied Sciences and Technology, https://doi.org/10.1007/978-981-19-7049-8_7

decisions in dynamic and complex situations under physical load is a major challenge for referees in team sports such as soccer. Moreover, frequent match schedules for the referees may bring about burnout thereby affecting their decision-making ability. Hence, it is important to examine the association between match loadings and decisions error in top European soccer league referees.

7.2 Clustering

A k-means clustering technique is considered for grouping the referees loading in this study. It is worth highlighting that a k-means assessment is a type of cluster analysis strategy that splits a set of data into k-predefined and non-overlapping subgroups termed clusters, with each data point assigned to just one of these groups [6–8]. The strategy attempts to connect as many inter-cluster data points as possible while keeping as many intra-cluster data points as distinct as possible. The cluster analysis was performed to categorize the number of matches officiated by the referees referred to as "match loading" and the frequency of decision errors during the matches. As a distance measure, Euclidean distance was employed to assign the formation of the two discovered clusters, namely high load referee (HLF) and low load referees (LLF). For the statistical analysis, the clusters were treated as dependent variables, while the decision errors committed by the referees were considered as independent variables [9–12].

7.3 Classification

In the present study, the artificial neural networks (ANN) model is employed to examine its capability in identifying the clustered classes, viz. HLF and LLF. A single hidden layer ANN topology was formulated, i.e. $6 - X - 1$. The number of hidden neurons, X, is varied between 10, 50, 100, 150, 200, 250, and 300, respectively. The activation function is also varied between ReLU, logistic as well as hyperbolic tan (tanh). The optimization algorithm in this study is set to Adam, while the maximum iteration is set to 1000. The optimized hyperparameters are identified by the mean classification accuracy (CA) attained on a combination of the hyperparameters through a technique known as grid search that performs a fivefold cross-validation method on the training dataset. It is worth noting here that initially the dataset was split into a 70:30 ratio for training and testing, respectively. The models were developed on Spyder IDE running on Python 3.7 with its associated scikit-learn libraries. The performance of the optimized model is deliberated via the CA, confusion matrix as well as the geometric mean (G-mean) score [13].

7.4 Results and Discussion

Figure 7.1 displays the grouping determined by *k*-means analysis in relation to the match loading and decisions error committed by the referees. The figure clearly shows that a distinct split was made between the HLF and LLF. In other words, the mean occurrences of decisions error are much larger in HLF referees compared to LLF.

Table 7.1 reveals the decision error indicators distinguishing the high and low load referees of European soccer leagues. It could be seen that out of the seven initially investigated decision error indicators, four were found to be statistically significant, i.e. decisions which led to disputes, foul not detected, offside not detected, as well as goal disallowed $p < 0.05$. Nonetheless, no statistically significant difference was observed between the two groups of referees in yellow cards and red cards not shown $p > 0.05$. This reflects that there is a strong connection between match loading and decision errors among the referees in elite soccer. Thus, there is sufficient evidence to postulate that high match loading could impair referees' decision-making in European soccer championships.

A total of 21 ANN models were developed and evaluated. From the sensitivity analysis carried out, the ANN model which utilized 50 hidden neurons on its hidden layer with the logistic activation function yielded a mean CA of 0.689 (±0.111) on the training dataset. Upon further inspection of its efficacy on the test dataset, it was further demonstrated that the optimized model could yield a CA of 66.85%. The confusion matrix of the test dataset is illustrated in Fig. 7.2. Nonetheless, a fair

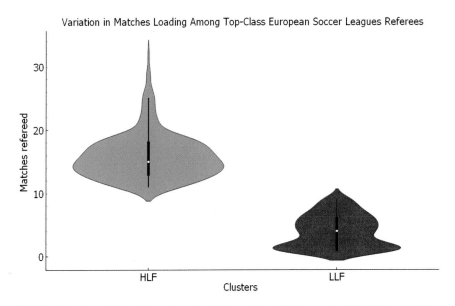

Fig. 7.1 Grouping of the referees based on their number of fixtures schedule. HLF = high load fixture, LLF = low load fixture

Table 7.1 Decisions error variations between referees of different match loadings

Decisions error indicators	Match load fixtures group mean (SD)		P-values
	HLF ($n = 326$)	LLF ($n = 276$)	
Decisions which led to disputes	1.721 (3.066)	0.605 (1.236)	0.0001
Foul not detected	0.623 (1.002)	0.167 (0.513)	0.0001
Offside not detected	0.334 (0.629)	0.149 (0.395)	0.0001
Yellow cards (not shown)	0.067 (0.335)	0.022 (0.146)	0.101
Red cards (not shown)	0.003 (0.055)	0.007 (0.085)	0.470
Goal disallowed	0.387 (0.739)	0.254 (0.592)	0.018

Note HLF—high load fixtures; LLF—low load fixtures

G-mean score of 0.6697 was obtained, suggesting a fairly balanced accuracy across the classes is attained.

The overall findings from this study demonstrated that there is a relationship between high match loading and decisions error in European soccer league referees. A number of decision error indicators that consist of decisions which led to disputes,

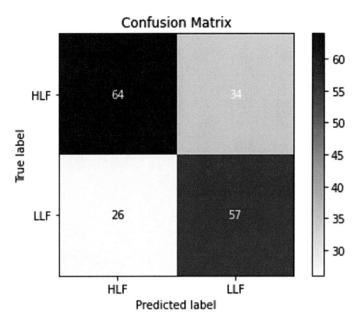

Fig. 7.2 Confusion matrix on the test dataset for predicting match loading and decisions error among the referees

foul not detected, offside not detected, as well as goal disallowed, are observed to be significant markers that differentiate the high and low load fixtures referees in their decision-making ability. In essence, the high load fixture referees committed more errors in the aforementioned indicators in comparison with the low load fixture referees. While there is existence of limited studies that investigate the association between the number of matches officiated and decision errors among European soccer referees, decisions and reasoning ability of soccer referees have been demonstrated to decline as the match progresses, i.e. from quarter two to three and from quarter three to four in a laboratory-based setting [14]. Moreover, previous studies indicate a weakening of reasoning and decision-making effectiveness in the second compared to the first half period of a soccer match under a high physical load [15, 16]. It has also been reported that referees' decisions are perceptual and cognitive processes that might be affected by a high level of physical load [17]. It is thus possible that physical load impairs referees' attentional control, thereby impairing their decision-making ability.

7.5 Summary

The present investigation revealed that there is a relationship between high match loading and decisions error among top European soccer league referees. A set of decision error indicators, namely decisions which led to disputes, foul not detected, offside not detected, as well as goal disallowed, are shown to be essential in differentiating the levels of errors committed between the high and low fixtures load referees. The high load fixture referees committed more errors in the said indicators in comparison with the low load fixture referees. Thus, it is recommended that fixtures related to loads of referees should be considered before the assignation of matches in European soccer leagues.

References

1. M. Methenitis, Laws of the game. Video Game Policy **1**, 11–26 (2019). https://doi.org/10.4324/9781315748825-2
2. A. Riiser, V. Andersen, A. Sæterbakken, E. Ylvisaker, V.F. Moe, Running performance and position is not related to decision-making accuracy in referees. Sports Med. Int. Open **3**, E66–E71 (2019). https://doi.org/10.1055/a-0958-8608
3. R.G. Vargas, J.A.U. Ramírez, D.R. Valverde, M.M. Thyssen, J.P. Ortega, External and internal load of Costa Rican handball referees according to sex and game periods. E-Balonmano.com Rev. Cienc. Deport. **17**, 153–162 (2021)
4. E. Ozaeta, U. Fernández-Lasa, I. Martínez-Aldama, R. Cayero, D. Castillo, Match physical and physiological response of amateur soccer referees: a comparison between halves and match periods. Int. J. Environ. Res. Public Health **19**, 1306 (2022)
5. S. D'Ottavio, C. Castagna, Analysis of match activities in elite soccer referees during actual match play. J. Strength Cond. Res. **15**, 167–171 (2001)

6. H. Azahari, H. Juahir, M.R. Abdullah, R.M. Musa, V. Eswaramoorthi, N. Alias, S.M. Mat-Rashid, N.A. Kosni, A.B.H.M. Maliki, N.B. Raj, A multivariate analysis of cardiopulmonary parameters in archery performance. Hum. Mov. **19**, 35–41 (2019). https://doi.org/10.5114/hm.2018.77322

7. M.R. Razali, N. Alias, A. Maliki, R.M. Musa, L.A. Kosni, H. Juahir, Unsupervised pattern recognition of physical fitness related performance parameters among Terengganu youth female field hockey players. Int. J. Adv. Sci. Eng. Inf. Technol. **7**, 100–105 (2017)

8. Z. Taha, M. Haque, R.M. Musa, M.R. Abdullah, A. Maliki, N. Alias, N.A. Kosni, Intelligent prediction of suitable physical characteristics toward archery performance using multivariate techniques. J. Glob. Pharma Technol. **9**, 44–52 (2009)

9. R. Muazu Musa, A.P.P. Abdul Majeed, M.R. Abdullah, A.F. Ab. Nasir, M.H. Arif Hassan, M.A. Mohd Razman, Technical and tactical performance indicators discriminating winning and losing team in elite Asian beach soccer tournament. PLoS ONE **14**, e0219138 (2019)

10. M.R. Abdullah, A. Maliki, R.M. Musa, N.A. Kosni, H. Juahir, S.B. Mohamed, Identification and comparative analysis of essential performance indicators in two levels of soccer expertise. Int. J. Adv. Sci. Eng. Inf. Technol. **7**, 305–314 (2017)

11. M.R. Abdullah, R.M. Musa, N.A. Kosni, A. Maliki, M. Haque, Profiling and distinction of specific skills related performance and fitness level between senior and junior Malaysian youth soccer players. Int. J. Pharm. Res. **8**, 64–71 (2016)

12. M.A. Gipit, M.R.A. Charles, R.M. Musa, N.A. Kosni, A.B.H.M. Maliki, The effectiveness of traditional games intervention programme in the improvement of form one school-age children's motor skills related performance components. Mov. Health Exerc. **6**, 157–169 (2017)

13. A. Kulkarni, D. Chong, F.A. Batarseh, Foundations of data imbalance and solutions for a data democracy, in *Data Democracy* (Elsevier, 2020), pp. 83–106

14. R.D. Samuel, Y. Galily, O. Guy, E. Sharoni, G. Tenenbaum, A decision-making simulator for soccer referees. Int. J. Sports Sci. Coach. **14**, 480–489 (2019). https://doi.org/10.1177/1747954119858696

15. J. Mallo, P.G. Frutos, D. Juárez, E. Navarro, Effect of positioning on the accuracy of decision making of association football top-class referees and assistant referees during competitive matches. J. Sports Sci. **30**, 1437–1445 (2012). https://doi.org/10.1080/02640414.2012.711485

16. H. Ahmed, G. Davison, D. Dixon, Analysis of activity patterns, physiological demands and decision-making performance of elite Futsal referees during matches. Int. J. Perform. Anal. Sport **17**, 737–751 (2017). https://doi.org/10.1080/24748668.2017.1399321

17. W.F. Helsen, C. MacMahon, J. Spitz, Decision making in match officials and judges, in *Anticipation and Decision Making in Sport* (Routledge, 2019), pp. 250–266. https://doi.org/10.4324/9781315146270-14

Chapter 8
Summary, Conclusion, Current Status, and Future Direction for Referees' Performance in European Soccer Leagues

8.1 Summary

It has been shown from the findings of the current investigation that several factors influence the decision-making ability of the referees officiating across different leagues in the European championship. It is demonstrated that various types of misconduct and foul-related activities, including fouls, challenges, challenges per foul, yellow cards, fouls per card, air challenges, ground challenges, risky play, misconduct, and attack wrecking, are common in the league. Similarly, numerous tactical and misconduct-related offences that comprised attack wrecking, ground challenges, misconduct, dangerous play, air challenges, and challenges off the ball are found to be non-trivial in predicting the likelihood for on-field referees to consult VAR. It has been identified from the findings of the current brief that the average referee's distance to the penalty box plus 2 m around fouls attracted more VAR assistance compared to the average referee's distance in other zones except for the penalty box plus 2 m around. Similarly, referees' decisions in key zones, distance 8–20 m, evoke more VAR assistance in comparison with the referees' key zones, distance to 8 m. Moreover, referees' decisions in other zones from a close range of less than 25 m are shown to invite higher VAR interventions when compared to the decision in key zones distances to 8 m and distances more than 20 m.

It has been revealed that there is a correlation between decisions error and specific times in match play. It is demonstrated that high-decisions error is mostly committed by the referees during a ball in play time, injury time, and possibly at the second half injury time as compared to first half injury time play. A set of event-related variables, namely foul not detected, offsides not detected, yellow cards not shown, red cards not shown, and goals disallowed, are shown to be essential in differentiating the referees in their ability to deliver the appropriate decisions during match play. It has been shown that referees of fixtures with high load committed more errors than those with low load.

R. Muazu Musa et al., *Data Mining and Machine Learning in High-Performance Sport*, SpringerBriefs in Applied Sciences and Technology, https://doi.org/10.1007/978-981-19-7049-8_8

8.2 Conclusion

At the highest performance level, soccer is defined by intense competitions, a high level of energy, technical, and tactical ability coupled with a lengthy period of play. Referees are expected to officiate the game and deliver undisputable decisions during match play. As a result of the increased pace and intensity of the game in recent times, the referees must adjust to and cope with the physiological and psychological demands inherent in the game. The referees are also required to make correct decisions and adhere to the game's laws and regulations during the game. These requirements and characteristics make referee work challenging. To mitigate these challenges, a video assistant referee (VAR) was introduced to support and improve the decision-making of on-field officials. However, despite the incorporation of VAR into the present officiating system, referee performance is yet to be error free.

Decision-making in dynamic and highly competitive situations under physical and physiological loads is a major challenge for referees in high-performance sports such as soccer. In most cases, referees are found lacking in anaerobic capacity, which is crucial as most of the important and decisive events in the game occur in the anaerobic zone. It is worth noting that the proximity of the referees to where an important event occurs is pivotal for making an accurate decision. Hence, to limit the probability of errors caused by these components of officiating, referees should undergo structured training to improve their functional skills, with a focus on boosting anaerobic capacity, under the supervision or guidance of a professional trainer.

In this brief, we explored the application of data mining and machine learning techniques in studying the activity pattern, decision-making skills, misconducts, and actions resulting in the intervention of VAR of the top-level European soccer league referees. A total of 6232 matches from five consecutive seasons officiated across the English Premier League, Spanish Laliga, Italian Serie A as well German Bundesliga were studied. It is also worth concluding that the application of data mining and machine learning techniques is vital for studying the activity pattern, decision-making skills, misconducts, and actions resulting in the intervention of VAR of the top-level European soccer leagues referees. It envisioned that the findings from the current brief could be useful in recognizing the activity pattern of top-class referees which is non-trivial for the stakeholders in devising strategies to further enhance the performances of referees as well as empower talent identification experts with pertinent information for mapping out future high-performance referees.

8.3 Current Status and Future Direction

Fans, players, and head coaches all around the world frequently complain about referees' uneven implementation of the rules as well as apparent prejudice against their side. Refereeing decisions may make or break a team's chances of winning a title, qualifying for lucrative European play, or escaping relegation. The adoption of VAR in major European soccer leagues has recorded some success thus far. There is an increase in respect among players through the reduction of misconducts such as verbal abuse and elbowing opponents' players. There is also an improvement in the issue of mistaken identity during card issuance where the VAR proved to help point out the appropriate player who deserved the punishment. However, there are many aspects where the VAR still fell short. For instance, the VAR technology is yet to solve the problem of fair play as players continue to dive (also known as simulation) for referees to award fouls to the players. There is also an issue of transparency as neither the fans, players nor coaches are cognizant of what mainly transpires between the referee and VAR during communication before reaching decisions. Consequently, there is still a need for improvement in said aspects. Moreover, referees are required to be more enlightened on how and when to use the available technology at their disposal for making more accurate decisions.

Since the initiation of the Goal Line Technology at the 2012 FIFA Club World, the usage of technology in soccer has increased tremendously. As a result, the International Football Association Board (IFAB) implemented the video assistant referee (VAR) system, as a second attempt for the European soccer governing body to make better judgements via the application of video technology. Moreover, FIFA recently declared that players' bodies would be monitored using artificial intelligence technology to make offside calls in the upcoming 2022 World Cup. It was disclosed that the semi-automated technology includes a sensor in the ball that sends its location on the field 500 times per second, as well as 12 tracking cameras positioned beneath stadium roofs that utilize machine learning to track 29 spots on players' bodies. The technology is envisioned to provide referees with more information to help them deliver more accurate decisions during match play. FIFA, on the other hand, asserted that the referees and assistant referees are still in charge of making decisions on the field of play. The success or failure of the system will still be subjected to evaluation upon completing the tournament. As such, more advanced statistical techniques will be required to identify the strength and weaknesses of the system. Machine learning coupled with data mining techniques is shown to be vital in providing insights from a large dataset which could be used to draw important inferences that can aid decision-making for diagnostics purposes and overall performance improvement. Thus, the application of ML in evaluating sporting performances is a non-trivial task. It is, therefore, envisaged that the technique employed in the current brief should be extended to other sports and physical activity-related domains.

Printed in the United States
by Baker & Taylor Publisher Services